T0312665

5G: 2020 and Beyond

RIVER PUBLISHERS SERIES IN COMMUNICATIONS

Consulting Series Editors

MARINA RUGGIERI
University of Roma "Tor Vergata"
Italy

HOMAYOUN NIKOOKAR
Delft University of Technology
The Netherlands

This series focuses on communications science and technology. This includes the theory and use of systems involving all terminals, computers, and information processors; wired and wireless networks; and network layouts, procontentsols, architectures, and implementations.

Furthermore, developments toward newmarket demands in systems, products, and technologies such as personal communications services, multimedia systems, enterprise networks, and optical communications systems.

- Wireless Communications
- Networks
- Security
- Antennas & Propagation
- Microwaves
- Software Defined Radio

For a list of other books in this series, visit www.riverpublishers.com
http://riverpublishers.com/river publisher/series.php?msg=Communications

5G: 2020 and Beyond

Ramjee Prasad
Founder Director,
Center for TeleInFrastruktur (CTIF),
Aalborg University,
Denmark

Founder Chairman,
Global ICT Standardization Forum for India (GISFI),
India

River Publishers
Aalborg

Published, sold and distributed by:
River Publishers
Niels Jernes Vej 10
9220 Aalborg Ø
Denmark

ISBN: 978-87-93237-13-1 (Print)
     978-87-93237-14-8 (Ebook)
©2014 River Publishers

Dedicated to

My grand children Sneha, Ruchika,
Akash, Arya and Ayush.

Contents

Preface

नहिकश्चित्क्षणमपिजातुतिष्ठत्यकर्मकृत्।
कार्यतेह्यवशःकर्मसर्वःप्रकृतिजैर्गुणैः ॥

Na hi kaścitkṣaṇamapi jātu tiṣṭhatyakarmakṛt।
Kāryate hyavaśaḥ karma sarvaḥ prakṛtijairguṇaiḥ॥॥

na... kaścid – Nobody; jātu – remains; tiṣṭhati – remains; a... kṛ t – without doing; karma – actions; api – even; kṣaṇam – for a moment; hi – because; sarvaḥ – every being; avaśaḥ — who has no free will; kāryate – is made to do; karma – action(s); guṇiḥ– by the qualities; jaiḥ – born; prakṛti – of Prakṛti

One cannot remain without engaging in activity at any time, even for a moment; certainly all living entities are helplessly compelled to action by the qualities endowed by material nature.

The Bhagavad Gita (3.5)

The foremost focus of this book is on next major phase in mobile telecommunication technology – fifth generation mobile technology (5G). The fifth generations of mobile systems will be a major stride in the mobile technology. Its standards would be beyond that of 4G. 5G is also referred as beyond 2020 mobile communications technology. Many scientific papers discuss the key concepts and techniques for the implementation of 5G systems. Some of them are, new data coding and modulation techniques, cognitive radio technology, multi-hop networks, Pervasive networks providing ubiquitous computing and so on. 5G can provide speed that would be at-least an order of magnitude higher than the mobile technology used these days.

Mobile giants such as Samsung, Ericsson, Huawei, Nokia Networks and Alcatel-Lucent have started their research on the development of 5G. Though some of the companies claim that they were successful in transmitting the data of 1.056 Gbps over a distance of 2 kilometres at extremely high frequencies, they do not expect these technologies to be commercialized till 2020.

The book discusses on providing the goals and implementation of a novel concept called Wireless Innovative System for Dynamically Operating Mega Communications (WISDOM) which becomes the main 5G definition point. Also, this book gives an idea of the sophisticated techniques to be implemented for making 5G a successful mobile generation than its predecessors. Certain contents for Chapter 2 are based on WISDOM project proposal, Chapter 3 is based on PhD thesis of Rajarshi Sanyal and Chapter 4 is based on Purnendu S.M. Tripathi's thesis.

The book is organized into the following chapters portrayed in the form a tree.

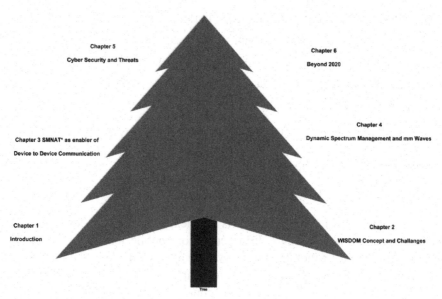

Figure 1 Structure of the book.

This book is an outcome of the recent effort in my academic garden, and I derive a sense of satisfaction from it similar to that drawn by a gardener on seeing blooming flowers. Further suggestions and comments to enhance the book and the concepts discussed therein are highly welcome.

List of Acronomys

1G	First Generation Mobile Technology
2G	Second Generation Mobile Technology
3G	Third Generation Mobile Technology
4G	Fourth Generation Mobile Technology
5G	Fifth Generation Mobile Technology
AMPS	Advanced Mobile Phone System
AP	Application Points
BMWA	Beam Width Multiple Access
CDMA	Code Division Multiple Access
CN	Cognitive Networks
CUS	Collective Use of Spectrum
CMOS	Complementary Metal Oxide Semiconductor Components
CONASENSE	COmmunication-NAvigation-SENsing-SErvices
CPS	Cyber Physical Structure
CR	Cognitive Radio
DoS	Denial of Service
DSSS	Direct-Sequence Spread Spectrum
EDGE	Enhanced Data Rate for GSM Evolution
FPGA	Field Programmable Gate Array
FDD	Frequency Division Duplex
FDMA	Frequency Division Multiple Access
GPRS	General Packet Radio Service
GSM	Global System for Mobile Communications
GIMCV	Global Information Multimedia Communication Village
GICT	Green Information Communication Technology
HAP	High Altitude Platform
HBC	Human Bond Communication
ITS	Intelligent Transport System

IDP	Intrusion Detection and Prevention
IoT	Internet of Things
LOS	Line of Sight
LTE	Long Term Evolution
M2M	Machine-to-Machine
MAC	Medium Access Control
MEA	Microwave Endometrial Ablation
MANET	Mobile Ad Hoc Networks
MIMO	Multiple-Input and Multiple-Output
NTT	Nippon Telephone and Telegraph
NMT	Nordic Mobile Telephones
OPEX	Operational Expenditure
OFDM	Ortho Frequency Division Multiplexing
P2M	Person-to-Machine
PHY	Layer Physical layer
QOS	Quality of Service
QoL	Quality of Life
RFID	Radio Frequency Identification
SDR	Software Defined Radio
TDMA	Time Division Multiple Access
TACS	Total Access Communication Systems
UMTS	Universal Mobile Telecommunication System
UE	User Equipment
VPN	Virtual Private Networks
V2I	Vehicular-to-Infrastructure
WISDOM	Wireless Innovative System for Dynamically Operating Mega Communications
WiMax	Wireless Interoperability for Microwave Access
WMAN	Wireless Metropolitan Area Network
WLAN	Wireless Local Area Network
WPAN	Wireless Personal Area Network
WSN	Wireless Sensor Networks
WWAN	Wireless Wide Area Network

List of Figures

List of Table

Acknowledgements

I would like to express my heartfelt gratitude to my colleagues and students contributing in developing the concept of 5G. Specifically, to Sahiti for the overall design and concept of the book; Rajarshi for providing inputs to Chapter 3; Purnendu for inputs to Chapter 4, and Prateek for inputs to Chapter 5 and in formatting the book. I am also thankful to the partners involved in the WISDOM proposal that led to formalization of the WISDOM architecture.

Finally, I would like to thank several of my colleagues and students who played a role in initiating 5G research at CTIF namely, Albena, Hanuma, Neeli, Rasmus, Sofoklis, Ambuj, and so on.

1

Introduction

In the last decade world has witnessed a tremendous growth in the wireless technologies. The result is an enormous growth in the evolution of various wireless communication devices like smart phones, laptops, tablets. Due to the enormous growth in the number of these devices for various applications, there is also the requirement for the development of various wireless solutions that are cost effective and reliable. The focus on the mobile wireless communications is dominant as it is the main means of communication in the world.

Based on the type of services and data transfer speeds, mobile wireless technologies have been classified according to their generations. They are shown in the form a tree in the Figure 1.1 for easy understanding and are explained below.

1.1 Mobile Wireless Technology Generations

The first generation mobile systems (1G) were analogue. The first cellular system was put into operation in 1978 by Nippon Telephone and Telegraph (NTT) in Tokyo, Japan. Some of the most popular analogue cellular systems were Nordic Mobile Telephones (NMT) and Total Access Communication Systems (TACS), Advanced Mobile Phone System (AMPS). AMPS and TACS use frequency modulation technique for the radio transmission. The inevitable disadvantage of this generation of mobile systems is that, even though these systems rendered handover and roaming capabilities the cellular networks were not able to interoperate among countries. Multiplexed traffic is carried over a Frequency Division Multiple Access (FDMA) system [2]. The second generation mobile systems (2G) are based on the standard Global System for Mobile Communications (GSM). GSM appeared first in 1991 and they are digital cellular systems. Digital communications enable advanced source coding techniques to be implemented thus allowing the spectrum to

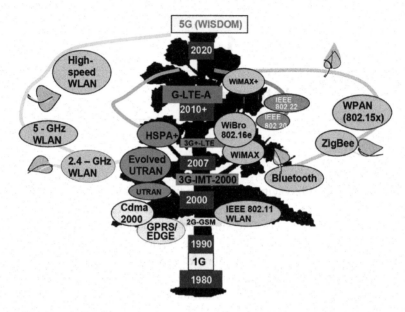

5G-WISDOM: Wireless Innovative System for Dynamically Operating Mega Communications 4G-LTE-A:Long Term Evolution Advanced   HSPA: High Speed Packet Access   3G+LTE:Long Term Evolution   3G-IMT-2000: International Mobile Telecommunications   EDGE: Enhanced Data Rate for GSM Evolution   2G-GSM:Global System for Mobile Communications 1G – NMT: Nordic Mobile Telephones TACS: Total Access Control System AMPS: Advanced Mobile Phone Systems

Figure 1.1   Tree of Standards [1].

be used much more efficiently. This also reduces the bandwidth required for voice and video [2, 3].

The main disadvantage with GSM is that it could handle a maximum data rate of 9.6 kbps which is too slow for the Internet related services. The 2G was subsequently superseded by 2.5G and 2.7G.

2.5G systems are evolved based on the General Packet Radio Service (GPRS) standard. These systems support Wireless Application Protocol, Multimedia Message Service, Short Message Service, mobile games, and search and directory mobile services.

2.75G systems are evolved based on standard Enhanced Data rate for GSM Evolution (EDGE) and this technology is an extended version of GSM. Data transfer rate is high compared to GPRS.

The third generation mobile systems (3G) systems were designed to provide a very high speed Internet access (about 384 kbps in burst mode). Some of the important services that these systems support are wide area wireless voice telephony, video calls, broadband wireless data and additional services like mobile television, Global Positioning System (GPS), other real time audio, video broadcast services. The three important technologies that paved the way to the development of 3G systems are given in reference [3]:

Universal Mobile Telecommunication System (UMTS) was developed that used Frequency Division Duplex (FDD) for forward and backward channels. With the help of wideband Code Division Multiple Access 5MHz channel spacing, high data rates of up to 2 Mbps can be achieved.

Time Division Synchronous CDMA (TD-SCDMA) used 1.6 MHz channel spacing.

The fourth generation mobile system (4G) offered very high speeds of up to 100 Mbps. The important feature of 4G systems are the high quality video and audio streaming over end to end Internet Protocol. The two important standards in 4G technologies are Worldwide Interoperability for microwave Access (WiMax) and Long Term Evolution (LTE). 4G is the current technology used all over the important places of the world. But there are many countries where the 4G services are not yet accessible because of the spectrum related issues [4, 5].

1.2 5G – From History to the Present and Future

The concept of realizing next generation communication systems in the form of 5th Generation communication network based on Wireless System for Dynamic Operating Mega Communications (WISDOM) followed by other leading initiatives at research facilities in industry and academia is as shown in Table 1.1. The operational concept of WISDOM has been discussed in detail in the following subsection and the various specific details related with its operation, and operational challenges that WISDOM – 5G would require to overcome have been elaborated in Chapter 2.

It has been long since the rollout of 4G based services by the cellular companies. Advent and possible utilization of 5G based services in future is already picking up. 5G based services are expected to commence from 2020. High data rate is expected to usher a new digital age ubiquitous communication for the masses that is unimaginable to be fulfilled by today's communication networks. 5G mobile wireless communications are expected to incorporate a large number of advanced technologies in order to increase the bandwidth further; Quality of Service (QoS), improve usability and security, decrease delays and cost of service. Some of the interesting services that users

Table 1.1   Significant 5G Initiatives till date.

Year	5G Initiative	Entity	Country
2008 February	WISDOM: Wireless Innovative System for Dynamic Operating Mega Communications [1]	Keynote Speech: **First International IEEE Conference on Cognitive Radio and Advanced Spectrum Management** Center for TeleInFrastruktur, Aalborg University	Denmark
2008 November	5G through WISDOM [7]	Center for TeleInFrastruktur, Aalborg University	Denmark
2008 November	5G systems based on Beam Division Multiple Access [7]	South Korea IT R&D department	South Korea
2012 May	First 5G System [7]	Samsung Electronics	South Korea
2012 October	5G Research Center [7]	University of Surrey	United Kingdom
2013 November	Research on 5G systems [7]	Huawei Technologies Co. Ltd	China

can experience are wearable or flexible mobile devices, Ultra High Display video streaming, smart navigation, mobile cloud, real time interactive games. Spectrum remains a key challenge for 5G, high frequency bands are to be explored to achieve those higher data rates than any other currently emerging technology. Some sources specify that when 5G arrives, it will have to handle billions of devices and myriad traffic types. It will offer improved reception and less network congestion, allowing for better connectivity and smoother roaming functionality [1].

1.3 WISDOM (Wireless Innovative System for Dynamically Operating Mega Communications)

WISDOM is an important and novel concept that defines 5G. The main reason for its description in this chapter is that it is an evolution towards a new communication connectivity era, which will offer the frequencies up to Tera Hertz and data rate up to Tera bps.

The fastest communication and ubiquitous connectivity is the foremost priority of the present era with the need of quick data transfer, distant business correspondence by sharing data and use of single IP for worldwide connectivity using hand held mobile.

To make it happen, the 5G network is assumed as the perfection level in mobile technology, which provides real life mobility. In this regard WISDOM is expected to advance the state-of-the-art in the architecture of next generation of wireless networks and cognitive technologies for higher data rates up to 1 Tera bps.

The relation between WISDOM and 5G can be expressed like this:

$$4G + WISDOM \triangleq 5G \text{ [1]}$$

The 4G and WISDOM concept will lead the wireless communication towards the realization of true 5G systems.

The main motivation for the development of the WISDOM concept for 5G is the needs of the society 2020 and beyond. These days, there is an exponential increase in wireless access bandwidth that is commercially available to the end user.

This will continue to increase in future wireless networks, taken by the rising needs of the mass market in the fields of bandwidth demanding applications such as entertainment, multimedia, Intelligent Transport Systems (ITS), telemedicine, emergency and safety/security applications [1].

Also the futuristic applications like 3D Internet, virtual and augmented reality that combines data for all senses, audio, visual, haptic, digital scent (e.g., tele-haptic applications, like planet or deep sea exploration), networked virtual reality (e.g., video streaming in social networks – users stream their own reality), and tele-presence (e.g., immersive environments with applications in both the commercial and military fields) can push the demand for real-time symmetric wireless connectivity to an individual with a data rate of 300 Mbps [1].

WISDOM aims for up to 1 Tbit/s (1 Tera bps = 10^{12} bits/s) wireless link rates in short-distance burst-mode or for up to 1 Tbit/s of system aggregated traffic with sustainable symmetric link rates of larger than 300 Mbps to mobile terminals at high speed [1, 3, 6, 7]. The communication at 1 Tbit/s is expected to be achieved utilizing millimetre waves (around 30 GHz and above), this has been discussed further in Chapter 4.

Existing wireless technologies like 3G+ cellular, WiMax IEEE 802.16e, Wi-Fi, Wi-Media as well as the corresponding emerging next generation networks (LTE/LTE-advanced, IEEE 802.16m, IEEE 802.11n, etc.) in the Wireless Wide Area Network (WWAN), Wireless Local Area Network

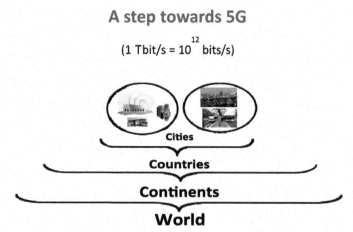

A step towards 5G

$$(1 \text{ Tbit/s} = 10^{12} \text{ bits/s})$$

Figure 1.2   Global Connectivity through WISDOM [1].

(WLAN) and Wireless Personal Area Network (WPAN) scales are not expected to meet such demanding needs for data rates.

WISDOM is a way towards 5G with a coverage extending from a city region to a country, continents and the world as shown in the Figure 1.2 forming a Global Information Multimedia Communication Village (GIMCV). Considering a user to shift between geographical locations swiftly and rely on different access networks, WISDOM based 5G network would be capable of dynamically aligning the communication network to support drastic changes in the overall network topology and therefore referred as dynamically operating mega communications based system.

In brief the key features of WISDOM are [6, 7]:

   a) combines established, competitive cellular standards with a promising frequency spectrum and novel enabling technologies;
   b) reduced coverage, electricity and operational expenditure (OPEX) costs;
   c) offers scalable and flexible technology options.

1.4 Global Information Multimedia Communication Village (GIMCV)

The different applications of the WISDOM network includes, home and office network, medical and health care, IT services, entertainment-movies high-speed data transfer, educational systems, rescue vehicular communications,

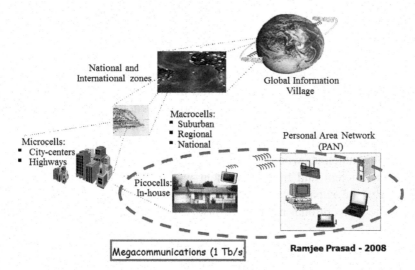

Figure 1.3 Global Information Multimedia Communication Village [1].

meteorology, security, aviation, global communication, smart town, virtual reality, intelligent transportation systems, and so forth.

All these diverse applications of WISDOM are the operating functionality and basis for the Global Information Multimedia Communication Village (GIMCV) as shown in the Figure 1.3.

The Global Information Multimedia Communication Village consists of national and international zones which are divided into macro cells (Suburban, Regional and National network). Macro cells comprise many micro cells (city – centres/highway networks), it further contains small pico-cells (in-house – network) where there are many small personal area networks. It gives scope for WISDOM application which varies from a person in home to globe from where they belong. It is the way of groupings of the devices in close vicinity of user [3].

The new 5G network is expected to improve the services and applications offered by GIMCV. WISDOM is a dynamic entity for human centric systems offering GIMCV.

1.5 Requirements of 5G

Based on the operational requirements of WISDOM based 5G to support a data rate of 1 Tbit/s there are certain things necessitated in the conventional networks. The requirement of the network is categorized into pico, micro and macro cells. It would be necessary for the mobile terminal to be recognized with a single ID so as to allow seamless network connectivity irrespective of the access network. The envisioned high data rate applications would necessitate that possible authentication and access validation to the mobile device are granted in miniscule time period, i.e., latency as low as 1 ms [8]. This would also necessitate minimization of possible shadowing effect and path loss due to be absolutely minimized. Utilization of distributed antenna systems (DAS) and multi-input and multi-output (MIMO) antennas would be unavoidable. The current MIMO systems would be insufficient and the requirement would be for massive MIMOs [9]. Similarly, the capability of mobile devices to communicate directly, bypassing the conventional network infrastructure, i.e., base station, would be necessary.

As stated earlier the high data rate operations would necessitate measures that can minimize latency to the absolute minimum. The capacity of mobile devices to initiate and establish cellular connections among themselves would be highly beneficial, commonly referred as device-to-device communication (D2D). Utility of D2D in WISDOM 5G communications has been elaborated extensively in Chapter 3. The current cellular operation spectrum bands are excessively crowded. The high data rate and existing spectrum situation would necessitate utilization of frequency bands that are conventionally not utilized for cellular radio communications. Frequency bands, higher than referred as mm bands mm-waves are capable of supporting the high data rated along with use of visible light communication (VLC), are two frequency bands that can suffice the operational requirements. Details about mm bands for WISDOM based 5G communication have been elaborated in Chapter 4. Apart from the using these frequency bands the core network would require to rely on cognitive radio technology for ensuring reliable high data services especially for supporting it on mobile devices that are mobile and change geographic locations swiftly. The relevant concepts for effective spectrum utilization in economic aspects of spectrum trading and sharing have been also elaborated in Chapter 4. The capacity of ubiquitous communication as envisioned to be provided in WISDOM 5G based communications would require specific measures in respect of security and privacy. Various aspects relating with security and privacy challenges and appropriate strategy to address them have been discussed in Chapter 5. Possible technologies such as Human Bond Communication and COmmunications-NAvigation-SENsing-SErvices that would be significant for fulfilling communication needs of society in 2020 in relation to 5G have been elaborated in Chapter 6.

1.6 Standardization of WISDOM

The major areas where the standardization of WISDOM shown in the Figure 1.4 is required are [3]:

- *Multimedia Communications* where it needs to focus on the areas of Machine-to-Machine (M2M) and Peer-to-Peer (P2P) with global identifications for home networking and smart cities and Techno-social Systems.
- *Cognitive Communications* where WISDOM based personalized cognitive communication includes all the educational, office, community, emergency, commercial and intelligent transportation systems.
- *Personalized Medicine* includes bioinformatics, multi-sensor networks, body sensors, and data protection and ethical guidelines.
- *Network without borders* basically comprises the wide range communications for the future Internet or the next generation networks. The main focus is on the Physical layer security, management and resource optimization, identity management, cooperative communications and Internet of things.
- *Embedded Optimal Resource and Computing* It has Energy harvesting techniques and models, time and power conscious hardware

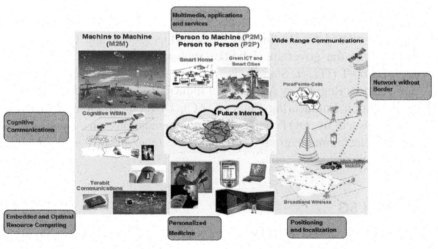

Figure 1.4 Standardization of WISDOM [1].

(HW)/software (SW) code sign methodologies, terminal energy trade-offs and energy aware reconfigurable and heterogeneous Architectures.

- *Positioning and Localization* includes the navigation systems, ubiquitous and cooperative localization, geo tagging, navigation ID systems and Robotics.

1.6.1 Global ICT Standardisation Forum for India

Global ICT Stadardisation Forum for India (GISFI, http://www.gisfi.org/) is playing a pivotal role in formalizing standards for 5G based on WISDOM that ensure the overall operational objective of uninterrupted Tera bps data rate to the end user and support human centric computing [13]. GISFI is addressing the aforesaid objective through its working groups that are looking into sub-aspects pertaining to it. There are seven working groups at GISI, which are as follows:

- Security and Privacy
- Future Radio Networks and 5G
- Internet of Things (IoT)
- Cloud and Service Oriented Network
- Green ICT

In addition to the working groups there are two additional groups:

- Special Interest Group
- Spectrum Group

GISFI has produced several documents covering the various aspects of standardization related to 5G through the working groups and they have been submitted to the government bodies such as International Telecommunication Union-T (ITU-T, Standardisation), Department of Telecom India (DoT) and Telecom Regulatory Authority of India (TRAI). Provided inputs to Telecommunication Engineering Center (TEC), DoT on "Implementations of directions issued by DOT on TRAI recommendations on 'Approach toward Green Telecommunications'" and emergency telecom services to TRAI amongst many others.

1.7 Vision of 5G

The overall vision of 5G can be summarized into the following broad also shown in Figure 1.5 [6].

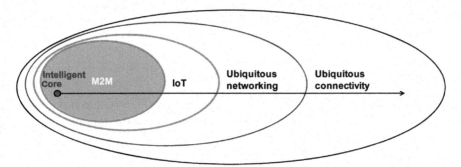

Figure 1.5 Plethora of technologies to deliver 5G services and applications [6], [13].

1.7.1 Enabling the 5G Intelligent Core

M2M and IoT are the two major criteria for realizing the intelligent core, which in turn is the key for enabling seamless ubiquitous networking and connectivity in a 5G context.

M2M and IoT are the key enabling technologies for a pervasive and always-connected 5G mobile services. Research challenges to fully deployable intelligent core are related but not limited to handling the big data collected through M2M and IoT communications (e.g., heterogeneous gateways, energy efficiency, decentralization of routing, naming and addressing), and to security, privacy and trust.

1.7.2 Enabling Ubiquitous Connectivity

This feature has two aspects. On one hand, technical challenges relate to sufficient coverage range even in a scenario of very high mobility and data rates, and on the other, to moving application from device-to-device without any content interruption.

Use of millimetre wave links novel multiple antenna concepts, virtualization, small cell deployments, and novel spectrum usage methods are some of the key research enabling areas for ubiquitous connectivity [1].

1.7.3 Enabling Ubiquitous Networking

This means that regardless of how many access networks are integrated for connectivity purposes, the quality of the delivered service must be retained end-to-end. In the aspect of increased importance of the cloud computing concept for supporting the big data originating from the intelligent 5G core, end-to-end ubiquitous networking will require interoperable decentralized service-oriented mechanisms with support of real-time interactions.

1.8 Summary

The evolution of mobile wireless communications from 1G to 4G has brought revolution in the communication among the people of the world. It is expected that 5G brings another revolution by offering very high data speeds. It incorporates many sophisticated technologies and uses important concepts like WISDOM for the better performance than their predecessors. This generation is expected to be rolled out by 2020.

References

[1] Ramjee. Prasad, (2008, February). Keynote Speech – Wireless Innovative System Dynamic Mega communications (WISDOM), in IEEE CogART'08: First IEEE International Workshop on Cognitive Radio and Advanced Spectrum,: http://www.wikicfp.com/cfp/servlet/event.showcf p?eventid=2104&copyownerid=538

[2] Ramjee Prasad, Universal Wireless Personal Communication, Artech House, 1998.

[3] Ramjee Prasad, Werner Mohr and Walter Konhäuser, Third Generation Mobile Communication Systems, Artech House, 2000.

[4] Shinsuke Hara and Ramjee Prasad, Multicarrier Techniques for 4G Mobile Communications, Artech House, 2003.

[5] R. Prasad and L. Munoz, WLANs and WPANs towards 4G Wireless, Artech House, 2003.

[6] Ramjee Prasad, (2008, November). Convergence towards Future, CTIF Workshop.

[7] 5th generation mobile networks, Wikipedia [online], http://en.wikipedia. org/wiki/5G

[8] Ramjee Prasad, Parag Pruthi, K. Ramareddy, The Top 10 List for Terabit Speed Wireless Personal Services, Wireless Personal Communication, vol. 49, no. 3, pp 299–309, 2009.

[9] Ramjee Prasad, Global ICT Standardisation Forum for India (GISFI) and 5G Standardization, Journal of ICT Standardization, volume 1-No.2, pp. 123–136, November 2013.

[10] Cornelia-Ionela, Neeli Prasad, Victor Croitory, Ramjee Prasad, 5G based on Cognitive Radio, Wireless Personal Communications, Volume 57, Issue 3, pp. 441–464, April 2011.

[11] 5G: A Technology Vision – Huawei [online], http://www.huawei.com/5g whitepaper/

[12] Cheng-Xiang et al., "Cellular architecture and key technologies for 5G wireless communication networks," Communications Magazine, IEEE , vol.52, no. 2, pp. 122,130, February 2014.

[13] Ramjee Prasad, Introducing 5G Standardisation, 11th GISFI Standard-isation Series Meeting, Bangalore, India, December 2013 [online], http://www.gisfi.org/news_events_details.php?id=73

2

WISDOM Concept and Challenges

The concept of WISDOM is evolved in order to meet the needs of interconnected society by offering ubiquitous terabit wireless connectivity. Based on the novel technologies, systems and network architecture WISDOM is a way towards 5th generation networks which will afford frequencies and data rates up to Tera Hertz and Tera bps, respectively.

WISDOM embodies the basic elements that are the building blocks of the future Internet innovations. The development and evolution of WISDOM offers a basic formula: $\mathbf{E} \sim \mathbf{MC}^5$

It is based on the combination of five independent vectors, Communication, Connectivity, Convergence, Content and Co-operation [1] as shown in Figure 2.1.

WISDOM is developed to enable the growth of an interconnected society, bridge the physical and virtual worlds by offering a seamless personalized rich digital experience for the end users, and also creating the optimal conditions for capitalizing on Future Internet innovations.

The top down design of WISDOM can be approached with the help of three founding pillars shown in Figure 2.2 which are [1]:

a) *Information theoretic performance/capacity estimation*:
   Different types of networking paradigms will lead the directions for engineering development.

b) *End-to-End performance optimization:*
   Protocol design for the end–to-end optimization is very efficient than the ones based on classical layered design.

c) *Cognitive networking principles.*
   Self-healing/self-organizing networks based on the above principle are essential to manage both, the complexity induced by a variety of possible usage scenarios and minimization of the spectrum and energy

15

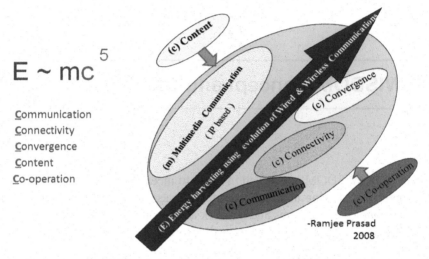

Figure 2.1   WISDOM a new paradigm [1].

Figure 2.2   Pillars of WISDOM [1].

requirements, in order to exploit the spectrum usage. WISDOM uses the concept of "cognitive network (CN)", which is a step ahead of Cognitive Radio.

2.1 WISDOM Objectives

WISDOM aims at enabling wireless infrastructure for the human-centric mega communications in 2020 and beyond. It aims at providing higher capacity and performance than any other current emerging technologies by [1]:

- Designing air interfaces and new systems that achieve a 3 to 5 times improvement over current wireless communications in terms of channel efficiency;

- exploiting larger channel bandwidths in uncontested areas of the spectrum in higher frequency bands and/or considering spectrum co-existence and sharing;
- employing smaller size cells and virtual cells with optimized dynamic spectrum management across different technologies;
- developing novel cross-layer and cross-network domain optimization technologies based on the principles of power efficient cognitive and cooperative communications;
- taking a comprehensive approach in developing a converged WISDOM system by jointly designing radio access systems and network protocols across a number of heterogeneous network architectures including ad-hoc, vehicular, mesh and next generation of cellular networks employing femto-cells and virtual cells efficiently connected to the wired core part of the Future Internet.

WISDOM will also help to design a *converged architecture* as shown in Figure 2.3 and network solution and evaluate its performance with the goal of enabling ubiquitous terabit wireless connectivity for human-centric

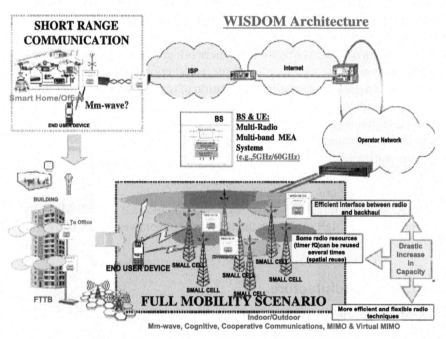

Figure 2.3 Converged architecture of WISDOM [2].

communications over the network of the future. WISDOM will design and develop wireless/wired interfaces and network protocols in order to integrate the wireless access networks to the fixed infrastructure as well as the core optical part of the Network of the Future. It means the focus of WISDOM will be on the design of wireless/wire line interfacing network nodes. These are the nodes that will facilitate the integration of wireless and wired (optical) network segments [2].

2.2 WISDOM System Requirements

In order to implement a WISDOM system the two main requirements are [2]:

- Mechanisms for enabling the transmission of Tera bps.
- Mechanisms for enabling mega communications.

The former can be achieved by the use of new physical layer techniques, such as new radio carrier-less transmission, to identify new spectrum bands, ultra high spectral efficiency mechanisms, MIMO and advanced physical layer (PHY) interfaces and advanced channel coding techniques. Also the new medium access control (MAC) and link layer are required for the strict QoS requirements that would allow the linking source and resource access for trillion device networks.

The latter can be achieved by the novel network protocols and architectures for heterogeneous networks such as femto cells, cooperative transmission, wireless-wired network integration, integration of high capacity satellite links and cognitive radio networks [1].

The 4G technologies together with WISDOM concept will show a way to the wireless communication to realize the true 5G systems.

To summarize [2]:

WISDOM communication interfaces are up to Tera bps link rate in a burst mode for short range communications.

WISDOM has a target of delivering a sustainable rate of 300 Mbps to mobile terminals at high speed for the needs of immersive applications and tele-presence on the move.

2.3 WISDOM Architecture

WISDOM architecture is shown in the Figure 2.4. It has three main components [2]:

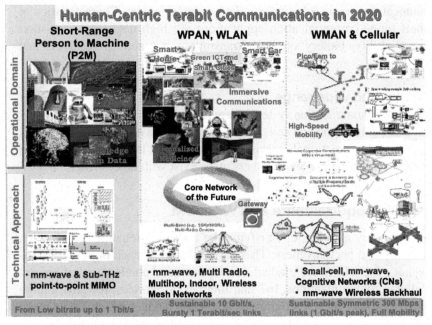

Figure 2.4 WISDOM Architecture [2].

A. Person to Machine (P2M)

The trend of person to machine communication (P2M) is inflaming the bandwidth demand and changes in communication because of its complexity, heterogeneity and integration of new systems and devices using the different network paradigms. These network paradigms consist of network devices ranging from computers, phones both landline and mobile, cameras, PDAs, collection of embedded machines like cars, washing machines, refrigerators, coffee vending machines, all are equipped with wired and wireless applications capabilities.

Applications of this P2M component are IoT, Wireless Sensor Networks (WSN), Mobile Ad Hoc Networks (MANET) etc. which transmits data from low bit rate up to Tera bps.

B. Shortrange

There are different short-range communication application such as smart homes, smart cars, Green Information Communication Technology (GICT) and smart offices, personalized medicines, Radio Frequency Identification (RFID), WLAN, security systems, keyless gates, future Internet, immersive communication, smart robotics, wireless accessories etc. Characteristics of the short range communication applications are high data rate, very short range, reliability, battery operated transmitters as well as receivers and low cost.

C. Cellular and wide range

The cellular communication system where in perfect cellular coverage occurs if each cell is a hexagon and cells can be arranged in a honeycomb. Most cell towers use Omni directional antennas to data transmit in a circular pattern. In some cases, cells overlap and in others, gaps exist with no coverage.

2.4 Users Requirements and WISDOM

Some of the key enablers of satisfying the user requirements are capacity, connectivity and pervasiveness. These drive the emergence of new environments that evolve from the gradual development and combination of present day cellular communications, IoT and Internet of Services, towards a more advanced vision of fully reprogrammable mobile devices which would make possible to communicate with each other autonomously based on a given event context and part of a scale-free self-organized communication system. Significant breakthroughs in the state-of-the-art are required to reach this level of performance paving us to what can be characterized as a new paradigm for future systems, namely WISDOM [1].

2.5 WISDOM Offerings

WISDOM intends to provide at least an order of magnitude of more capacity than any currently planned wireless radio technology. Creating wireless infrastructure that enables human-centric mega communications is the main aim of WISDOM.

WISDOM gives a novel cross-layer and cross-network-domain optimization technologies based on the principles of power efficient cognitive and cooperative communications. It also offers a design of new air interfaces and systems that achieve a 3–5 times improvement over current wireless communications in terms of channel efficiency. It helps in the exploitation of a large channel bandwidth in uncontested areas of the spectrum in higher frequency bands encouraging the techniques like spectrum co-existence and sharing. WISDOM development offers large evolution [2]:

- Making extensive use of small size cells and the concept of virtual cells, i.e., grouping a number of small cells in a synchronized manner in a way that they all operate in the same channel and are seen by the terminal as a single base station.
- Sharing of knowledge which is used to improve user safety, through intelligent rich presence and collaboration.
- Services for accessibility of users and knowledge based services into workflow and applications, regardless of device.
- Freedom for users to work from home, office or other location using high speed Internet connections.
- A logic based on dependability of services for modularization and locality which operates on "best effort" of technology usage.

2.6 WISDOM Impact

It is already mentioned that WISDOM affords terabit communications with a coverage extending from a city region, to a country, the continents, and the world. This is possible by combining personal- and cognitive radio-networks which is the basic operational concept of WISDOM.

WISDOM enables to capitalize on major innovations towards the future smart infrastructure by integrating under one interoperable umbrella leading technologies, such as advanced M2M communication technologies, autonomic networking technologies, data mining and decision-making technologies, security and privacy protection technologies, cloud computing technologies, with advanced sensing and actuating technologies.

2.7 WISDOM Challenges

The four main operational domains of WISDOM are:

1. Short range low mobility communications
2. Outdoor/Indoor cellular communications with full mobility
3. Converged Architecture
4. Security

The challenges faced by the WISDOM in the above domains are as follows:

Short range low mobility communications

The main target of WISDOM is to provide high data rates of about 1 Tera bps in short range communications. Some of the applications like

smart phones, smart cars, personalized medicines, WLAN security systems, Radio Frequency Identification (RFID), keyless gates, future Internet,
robotics, immersive applications, Green Information Communication Technology (GICT) and smart offices other wireless accessories etc. fall under this
category.

The key challenges to achieve high data rates in short range distances
are [1]:

a) Exploiting the particular bands in the spectrum range from the Extra
   High Frequency band, that allow for the design of systems that yield
   high data rate and significant bandwidth efficiency, as required for
   Tbit/s communications. Band frequency ranging from 70 GHz up to
   300 GHz should be considered in particular. Novel and unconventional
   solutions, both for RF and baseband design should be considered as the
   traditional design approaches put some limitations on the achievement of
   1-Tera bps connectivity. But, the high absorption rate at those high carrier
   frequencies poses great challenges for their utilization in Non Line of
   Sight (NLoS) and mobile connections.

   Exploitation of such high frequency bands opens many challenges,
   among them are:

   - available technology support;
   - channel characterization at those frequencies is lacking;
   - design of robust modulation and transmission techniques has to
     be done considering Complementary Metal Oxide Semiconductor
     (CMOS) components limitations;
   - pass band of digital device does not allow to use one channel with
     a bandwidth higher than 2 GHz today.

b) WISDOM short-range communications are based on multiple directional
   antennas transmitting to the same terminal in order to create spatial diversity and mitigate Line of Sight (LoS) blocking. Therefore, WISDOM
   requires novel PHY techniques and also the advancements towards the
   cognitive network architecture. Progress towards this CN concept can be
   done by carrying out research in a series of interrelated fields such as,
   directional links, adaptive modulation and coding, medium independent
   handover, cognitive radio and cooperative techniques at different layers
   of the protocol stack.

c) *The human-centric paradigm*: This requires a huge interaction of the user
   with her/his environment in order to interchange information related to
   the context, profile, role and other relevant information which in general

may help to optimize the whole network behaviour as well as the user perception.

These are the challenges the WISDOM should consider for achieving high data rates in short-range communications.

The scenario of network architecture for short-range communications is shown in the Figure 2.5. It consists of wireless communication systems components such as like CN, WSN, millimetre-wave (mm-wave) cooperative communications, MIMO and virtual MIMO, multi-radio and multi-band devices, multiple frequency bands and air interfaces [2].

There are:

- In mm-wave Communications: MIMO and Virtual MIMO links in the mm-wave bands Different network architectures (Scatter-nets, Multi-hop Mesh, ad-hoc, indoor wireless-wireline connectivity).
- Application points (AP) and User equipment (UE): it consists of mainly three types of wireless devices such as Multi-band, Multi-Radio, and Microwave Endometrial Ablation (MEA).

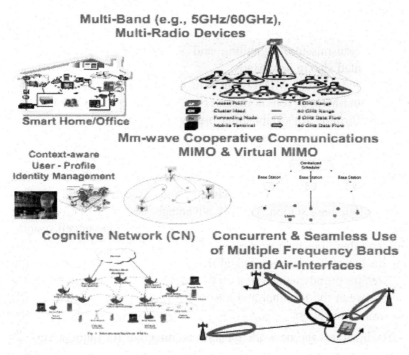

Figure 2.5 Network architecture scenarios for short-range communications [2].

- Different network architectures (Scatter-nets, Multi-hop Mesh, ad-hoc, indoor wireless-wireline connectivity).
- The concept of a CN: it is the combination of computer networks, machine learning, knowledge representation and network management. Cognitive network is a cognitive process which has end-to-end goal which is achieved by following the different network condition, planning, decisions for the different network condition with consequences of the decisions.
- The human-centric paradigm: these devices are configured in such a way that it will give paradigm which will be human centric, e.g., smart homes, offices, cars etc.

Outdoor mobile communications with full mobility

WISDOM aims at delivering a sustainable rate of 300 Mbps to an individual user at full mobility with a peak rate in excess of 1 Gbps, while the vision of the 1 Tera bps rate corresponds to aggregate capacity of a large number of users served in a metropolitan area. This target calls for a revolutionary step for the techniques to be adopted [2]:

a) WISDOM shall consider new communication systems, based on the suitable transmission, signalling and modulation techniques, to be implemented also in new bands.

b) MIMO and Virtual MIMO links in the mm-wave bands: MIMO and Virtual-MIMO (on both sides of the link) and various types of beam-forming for establishing highly directional links at high frequencies and in fast-fading environments.

c) Use of small cells and virtual cells

- A virtual cell is a group of small cells synchronized in a way that they all operate in the same channel and are seen by the terminal as a single base station. The key advantage of such a solution, is that mobility signalling is not increased compared to larger cell systems and therefore management overhead and terminal complexity are in the same order of magnitude. At the same time the solution retains the advantages of a small cell system since the short distance between the terminal and the nearest cell allows high bandwidth communication.

Figure 2.6 illustrates the network scenario/architecture for outdoor cellular system with the following listed components [1]:

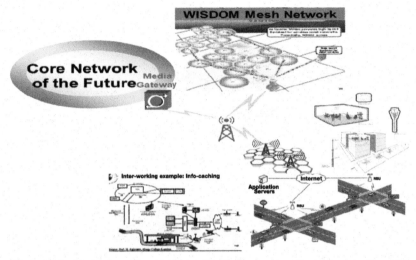

Figure 2.6 Network architecture scenarios for outdoor mobile communications [2].

- WISDOM Base Station and User Equipment: Multi-band, Multi-Radio, MEA Systems
- In mm-wave Communications
- MIMO and Virtual MIMO links in the mm-wave bands. Different network architectures (Multi-hop Mesh, ad-hoc, Wireless/Wireline backhauling, Intelligent Transport System (ITS)/Vehicular-to-Infrastructure (V2I))
- The concept of CN
- The human-centric paradigm

Figure 2.7 explains the network scenario for next generation cellular networks and it consists of network of radio cells. It is aimed to achieve Wireless Metropolitan Area Network (WMAN). At the WMAN scale, the vision of Tera bps wireless corresponds to aggregate capacity of all wireless users served in a metropolitan area. For example, 400 radio cells in a metropolitan city, each cell serving 10 users, using 250 Mbps each for virtual reality applications: 400 x 10 x 250 Mbps=1 Tera bps [1].

Converged Architecture

WISDOM will help to design a converged architecture and network solution with the goal of enabling ubiquitous terabit wireless connectivity for human-centric mega communications over the network of the future.

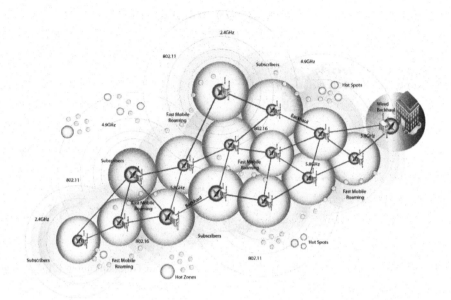

Figure 2.7   Scenario for next generation cellular networks [2].

WISDOM will design and develop wireless/wired interfaces and network protocols in order to integrate the wireless access networks to the fixed infrastructure as well as the core optical part of the Network of the Future. For an optimized wireless/wire line network WISDOM will propose some novel solutions addressing the overall network architecture and the interconnection of the different technology parts of the network with the special focus on interfacing network nodes. This is because these nodes will play the role of edge devices as far as the optical network is concerned and will provide aggregation, traffic shaping and traffic engineering capabilities. For end–to-end user optimization WISDOM will design, developand evaluate novel transport protocols compatible to the wired part (optical fibre) of the network that are able to support the characteristics of both the metro and the core part of the network offering abundant bandwidth in an efficient and cost effective manner. The converged WISDOM architecture will comprise a number of self-organizing/self-healing wireless access networks whose design is based on cognitive and cooperation networking principles. The design of these access networks is based on cognitive networking principles.

The key building blocks of the WISDOM access networks are the WISDOM-node and the WISDOM-station, as shown in Figure 2.8. The

WISDOM-node is envisioned as an intelligent cognitive multi-radio, multi-band, MIMO mobile device capable of operating in a variety of spectrum allocation and interference conditions by selecting cross-layer, cross-network optimized physical and network layer parameters often in collaboration with other radios even if they belong to different co-located networks. The *WISDOM-station* is envisioned as fixed or mobile WISDOM-node that is chartered with the task of providing seamless interconnection between the wire line and wireless parts of the converged architecture [2].

WISDOM node has a vital role in the design of the network algorithms and protocols at the local network and global internetworking levels. Specifically, at the local level, support for cross-layer and cross-network optimization algorithms in autonomous cognitive networks requires an advanced distributed control and management framework. At the global internetworking level, clusters of cognitive networks represent a new category of access networks that need to be interfaced efficiently with the wired network infrastructure both interms of control and data.

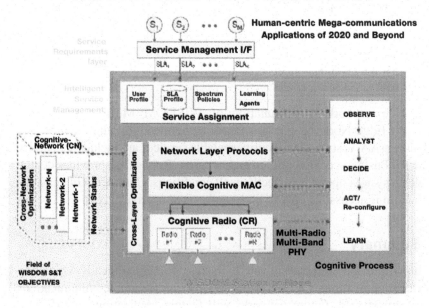

Figure 2.8 WISDOM Node and WISDOM Station [2].

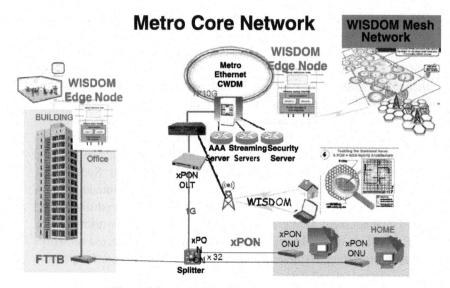

Figure 2.9   End–to-End System Architecture [2]

End-to-End system

The end-to-end system architecture shown in Figure 2.9 consists of the following components [1]:

a) WISDOM Mesh Network
b) WISDOM Edge Node

Metro core network is permutation of WISDOM Mesh Network and WISDOM Edge Node. In this mishmash authentication, authorization and accounting (AAA) streaming security scalable services are included. Security includes robustness, confidentiality and integrity. Robustness provides degree to which a system operates perfectly in all conditions. Confidentiality ensures information is accessible only to those authorized to have access.

2.8 Summary

The aim of the Wireless Innovative System for Dynamically Operating Mega-communications (WISDOM) is to provide terabit communications to the future world. Future networks will feature applications and devices that are highly personalized and humancentric. WISDOM aims to support interest-based service provisioning, where an interest may be based on user-behaviour or user profile. It will design and develop technologies, systems and network architectures that will enable ubiquitous 1 Tb/s wireless connectivity and communication with coverage extending from C^3W through C^5 (C^3 stands for City, country, continent and W stands for World; C^5 represents Communication, Connectivity, Convergence, Content, and Co-operation) [1]. The important point is that WISDOM combines personal- and cognitive radio-networks towards robust and efficient future networks that will enable a smooth shift from static to dynamic communications, successful businesses that are financially viable and provide secure transfer of information.

References

[1] Ramjee Prasad, (2008, February). Keynote Speech – Wireless Innovative System Dynamic Megacommunications (WISDOM), in IEEE CogART'08: First IEEE International Workshop on Cognitive Radio and Advanced Spectrum,: http://www.wikicfp.com/cfp/servlet/event. showcfp? eventid=2104&copyownerid=538
[2] Ramjee Prasad. (2008, November). Convergence towards Future, CTIF Workshop.

3

SMNAT as Enabler of Device-to-Device Communication

D2D communication for WISDOM based 5G would be a significant aspect, especially as a mechanism to minimize the latency incurred for accessing a network that would be unacceptable for the WISDOM based 5G. Smart Mobile Network Access Topology (SMNAT), an enabler for D2D in WISDOM, has been discussed in this chapter.

The 5th and 6th generation of mobile networks are envisioned to realize the bandwidth required for WWWW (Wireless Worldwide Wide Web). They should be highly dynamic in nature and actuate self-optimization of resources to support the bandwidth for ever increasing customer base. Spectrum crunch is one of the major issues. Hence these networks should thrive to achieve very high spectral efficiency and less Carrier to Interference ratio (C/I), relative to the existing technologies like LTE and WIMAX. Data speed per user should also not be compromised. Ubiquitous, instantaneous and always connected are some of the key prerequisites of a modern mobile device. It has been partly complied by LTE and some Network based applications based on the IMS (IP Multimedia Subsystem) [1] framework like Rich Communication Suite (RCS) [2]. The data requirement per user is facing an exponential growth with the high proliferation of the applications on the UEs available today. M2M and IoTs today play an important role in shaping up the data traffic patterns which can be sporadic or distributed over time. As the new device technologies usher in, the data throughput requirement per user soars. The smart phone and tablets with three dimension (3D) screens and cameras will be bandwidth hungry. Screen resolution will further increase leading to higher data rates. Online multiplayer gaming applications running on 5G gaming consoles with 3D displays will require higher throughput both for the uplink and downlink. Ultra High Definition (UHD) voice and UHD video realized through Rich Communication Services (RCS) [2] will contribute to the growth of data traffic. More dedicated data bearers will be required at the access and

31

core to guarantee the QoS promised to the user. Due to the swarm of M2M and IoT devices in the network, the mobility management related processes will be more intense and will consume more network resources. The nano nature of the devices with limited RF power will imply additional small-cells/microcells which will increase the intricacy of the access and core network.

The key question is how can one realize a network that can achieve the following:

1. Cater the bandwidth requirement of next generation mobile devices meant for human users as well for machines.
2. High Data rate over large coverage areas and dense demographies.
3. Reduce cost of network infrastructure and operation
4. Meet the bandwidth, latency, QoS requirement for supporting the next generation network and device applications.
5. Reduce power consumption in the network and the device
6. Lessen the complexity in the Access (radio) and core network layers

Living through the advancements in mobile network technology from 2G/3G–UMTS towards LTE (4G), one may notice that the primary emphasis is to enable higher bandwidth per user required to drive the next generation network and device applications. With the evolution of the mobile networks from 2G to 4G [3], there is an attempt to inject more symbols (data) per unit time. This has been made possible by the evolution of the modulation schemes from Binary Phase Shift Keying in case of 2G, to 64 Quadrature Amplitude Modulation for LTE-Advanced. In the current Mobile Radio Access Network there is a host of network processes, tightly synchronized and orchestrated by intelligent network elements, such as:

- Handover management
- Location management
- Call drop off management
- Interoperability and downward compatibility management
- Service control (like roaming control)
- Feature management

These processes involve complex signalling operations across the radio and the core network. The attempt to simplify the network in IMS [1] and LTE [4] is focused to make the core and access networks all IP. Following the evolution trail from 2G to 4G, a significant philosophical drift towards simplification in terms of the service logic related to mobility and location management is not discerned. A similar process for handover, frequency reuse, location updates and cancellations in 4G as compared with 2G [3] is witnessed. Hence the network elements of the state of the art still need to be equipped with the intelligence and processing power to handle all these complex signalling operations. This has a knock-on effect on the handsets too. It involves considerable amount of processing power, thus leading to more battery drainage. So it is imperative to look beyond the state of the art and conceive a technology which is devoid of these constrains and is ideal for the next generation mobile devices. The same goes for core network design. Mobility and Location management in the core network necessitates a substantial volume of signalling interaction between the HLR/HSS (Home Location Register/Home Subscription Service) and the MME (Mobility Management Entity) in LTE core [4] and MSC/VLR (Mobile Switching Center/Visited Location Register) in UMTS Core. Inefficient and un-optimized engineering and design of these network elements impact the service assurance, and finally degrade to the overall QoS offered by the network. Also, a substantial amount of resources is used up for addressing the wireless node in the cellular network, both from the perspective of available bandwidth and also the associated processing activities carried out in the different network elements at the Radio Access Network (RAN) and Core Network (CN).

Managing a mobile network is hugely resource and knowledge intensive, primarily because of the inherent complexity of the network architectures. For example, during cell planning one needs to consider the traffic demand to cover a specific region, availability of base station sites, available channel capacity at each base station, and the service quality at various potential traffic demand areas. The allocation of the right frequency at the cells to get an optimum frequency reuse factor is crucial to achieve smooth handovers, avoidance of call drops during handover, and overall elevation of the performance of the network. These are part of the overall cell planning activity and it requires a lot of resources, in terms of FTE (Full Time Equivalent) and hence it increases the operational cost. The Switching (Time/Space) matrix of a Mobile Switching Center of a 2G/3G network has a finite limit to make and break the number of calls. The Busy Hour Call Attempt handling capacity of a Switch depends much on engineering and dimensioning of this matrix. With the CSCF (Call Session Control Function) for IMS (IP Multimedia Subsystem) [1] and the MMEs (Mobility Management Entity) for the LTE [4], one does not need the Time and Space switching Matrix. The CSCF/MME acts more as SIP (Session Initiation protocol) router for IMS and SIP/GTP/DIAMETER router in case of LTE. However, for routing a call, these network elements still need to involve directly the Layer 7, which utilizes lot of resources and processing power of these network elements. This in turn makes the network elements expensive and increases the capital and operational cost of the mobile network. As a consequence of the convergence of heterogeneous mobile applications catered by the expanse of the mobile devices preordained for diverse deployment scenarios, the signalling inter-processes between the network and the user equipment become more intricate.

3.1 5G Communication Landscape

With the present day device capabilities, the mobile devices are in constant interaction with each other. The human-centric mobile device interacts with various wearable devices, like smart watches, wearable computers (e.g., Sixth Sense), SOS devices and health equipments. As the user moves around, these devices become more dynamic in nature. The devices, categorized as M2M or IoTs, may be as below (though this is an example and not an exhaustive list):

1. Environment sensors
2. Connected cars
3. Smart objects and robots
4. Health equipments
5. Small cells not owned by the mobile operators

According to the newly formed group called METIS (Mobile and wireless communications Enablers for the Twenty-twenty Information Society), which is a consortium of mobile Original Equipment Manufacturer, operators, academic institutions and automotive companies a 5G Network should have 1000 times higher mobile data rate volume per area, 10 to 100 times more connected devices, 10 to 100 times higher typical user data rate, 10 times longer battery life for low power devices and 5 times smaller end-to-end latency. The main objective of the project is to lay the foundation of 5G, the next generation mobile and wireless communications system. The aim is to let people seamlessly bridge the virtual and physical worlds offering the same level of all-senses, context-based, rich communication experience over fixed and wireless networks. Apart from these basic needs, one need systems which can work with the same performance level in the crowds. It should offer the same QoS and throughput as in office, or home as on the move. It should render low end-to-end latency and reliability to enable machine type applications.

3.2 Related Work on 5G

To meet these demands and to conceive a network which incorporates all these enhancements over the 4G, one may need to analyse the potential of new access technologies and architectural concepts and then identify the ones which can be the eligible contenders to serve as 5G technology

Some of them are as follows:

a) Multi-tier 5G Networks
b) PHY layer based Network Coding
c) Generalized frequency division multiplexing (GFDM)
d) Non Orthogonal Multiple Access (NOMA)
e) Cognitive Radio
f) D2D communication
g) Beam forming technology
h) In mm-wave technology
i) Massive MIMO

However, the scope of this chapter is not to investigate on all these options and not to provide a comparative analysis. Rather, some new paradigms which have the potential to play pivotal roles in shaping up the 5G network architecture and the D2D Communication Framework are staged. In D2D communication, user equipments (UEs) exchange information among themselves peer-to-peer over a direct link using the cellular resources instead of that through the base station or eNodeB. This is markedly different from small cell (femto cell) communication where UEs communicate with the help of small low-power cellular base stations. D2D users communicate directly while remaining controlled under the Cellular Access network. This optimizes resource utilization in a cellular network and boosts spectral efficiency.

This chapter will primarily delve in the following domains.

a) D2D communication: a study on the existing approaches with the new approach, the Smart Mobile Network Access Topology (SMNAT).
b) Integration of SMNAT with LTE-Advanced and 5G Core.
c) Security aspects in light of cooperative communication between two devices.

3.3 Cellular Device-to-Device Communication

Cellular D2D communication is meant to reduce the cellular traffic load by actuating a breakout from the UE itself and directly establish traffic towards the other paired UE using the cellular channel. As a collateral impact, this can significantly contribute to slacken the processor intensive signalling process as explained above. It leverages the benefit from the proximity between two devices and increases the overall resource utilization of the cellular network.

But it is imperative that one needs to come up with new methodologies for device discovery and pairing. Direct D2D technologies have already been developed in several wireless standards, aiming to meet the needs for efficient local data transmission required by variant services in personal, public and industrial areas. Some of the existing contenders are Bluetooth, Zigbee and direct Wi-Fi. With D2D communication, the aim is to find a method which is tightly integrated with the cellular network and uses the same spectrum as cellular operations.

Imparting D2D capability in a mobile device impacts the whole of the network framework and is not an inconsequential addition. Issues like authentication, real time billing, fraud control will crop up and the devices will directly interact with each other bypassing the network.

But on a positive note, the benefits are bountiful.

1. As the paired devices are in the same cell sector, or in the same cell (different sector) or in adjacent cells in physical proximity, high data rates with low latency can be achieved.
2. Depending on the proximity of the devices, the radio power level will be reduced. This will result in better battery life.
3. As the same radio resources are used for cellular and D2D communication, hence the average frequency reuse factor will be better.
4. For a traditional cellular network, one needs two distinct channels for uplink and downlink between the UE and the base station (BS). But in case of D2D communication, a single channel can be used for both directions. Hence the overall spectral efficiency is better.
5. The spectral efficiency can be further enhanced if the cognitive radio communication is used wherein the unused spectrum holes may be utilized for establishing direct communication between the two nodes.
6. As the D2D communications has limited dependence on the network infrastructure the devices could be used for instant communications between a number of devices within a range.
7. D2D on 5G would use licensed spectrum and this would enable the frequencies to be used to be less impacted by interference.
8. In times of natural calamities where some essential components of core and access network have failed, D2D communication can ensue.

A practical implementation scenario of an Over the Top (OTT) application meant from smartphones (for D2D communication) is from Google. A new mobile messaging application called FireChat is empowering nearby smartphone users to stay in touch even when there's no cellular service or

Internet connection. The messaging application harnesses a technology called wireless mesh networking, which might someday allow a myriad of devices to connect like links in a chain. The technique might someday be used to tie together thousands of devices with built-in radios and make it possible to be online without having to pay for the access. It could also enable online communications in remote areas or disaster zones without Wi-Fi or cellular signals.

D2D has been proposed as a Rel.12 3GPP feature. D2D Study Item had an approval in 3GPP SA1 (Services working group) in 2011, called ProSe (Proximity based Services) [5] which identifies the use cases and envisage the requirements including network operator control, security, Authentication, Authorization and Accounting (AAA), regulatory aspects, public safety requirements, integration with current infrastructure, network offloading. The ongoing discussion by ProSe includes evaluation requirements, D2D channel model, resource use, ProSe discovery and ProSe communication, etc.

ProSe Communication between two UEs in proximity is established by means of a communication path established between the UEs.

- The ProSe Communication path is established

  - by Direct communication between the UEs
  - or routed via the local eNB
  - by ProSe Discovery.

- Communication Process identifies that a UE is in proximity of another UE.

  - by Open [ProSe] Discovery

- ProSe Discovery without permission from the UE being discovered.

  - by Restricted [ProSe] discovery
  - ProSe Discovery that only takes place with permission from the UE being discovered.

D2D communication for M2M type devices is a topic of interest for many telecom researchers. This is because of the huge volume of the devices which may eventually clog up the Mobile network and jeopardize the human-to-human services. The study on D2D requirements for MTC Device to MTC Device scenarios covers

- The identification and functionality needed to set up a connection towards a MTC Device.
- The IMS domain may provide a solution for this required functionality.
- MTC Devices often act as a gateway for a MTC capillary network of other MTC Devices or non-3GPP devices

D2D group for Machine type communications (M2M) Study on Enhancements for MTC is a 3GPP specification – TR 22.888 [6] has been crafted in purview of these requirements.

As depicted in Figure 3.1, the scenarios that are covered are

- Devices Communicating directly
- Devices Communicating via MTC Server
- Devices Communicating with assistance from a Name resolution server

D2D Communication can be broadly categorized as:

1. Out of band
2. In-band

Out of band D2D essentially implies that the devices use a radio technology in the standalone mode to actuate communication with a paired device. The pairing method is in the purview of that specific radio technology.

Some existing out of band D2D communication methods are:

- Bluetooth
- ZigBee
- Near Field Communication
- Direct Wi-Fi

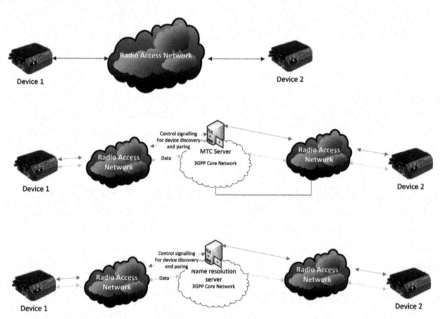

Figure 3.1   Three methods of D2D communication for M2M devices.

In-band D2D communication implies that the devices use the cellular spectrum and the cellular technology.

In this chapter a new concept for the realization of in-band D2D not by using new resource allocation methodologies but by using a new mobility management and addressing concept called SMNAT (Smart Mobile Network Access Topology) [7–13] has been presented.

The subsequent Sections 3.4–3.6 will bring out the various radio resource management technologies that can be employed in the in band D2D communication, as seen in the state of the art. These technologies are based on the underlay cellular network, hence they fall in the category of in-band D2D communication methods.

3.4 D2D Using Physical Layer Network Coding

In this technique [20], there are two cooperating mobile nodes which can relay network codes over the channel codes in two different paths, as shown in Figure 3.2. First it is being sent to the candidate mobile, and subsequently

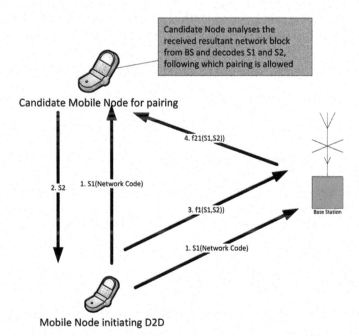

Candidate Node analyses the received resultant network block from BS and decodes S1 and S2, following which pairing is allowed

Candidate Mobile Node for pairing

4. f21(S1,S2))

2. S2   1. S1(Network Code)

3. f1(S1,S2))

Base Station

1. S1(Network Code)

Mobile Node initiating D2D

Figure 3.2 Interaction between the initiating device, candidate (pair) device and base station for realizing network assisted D2D communication.

it is being sent to the BS. The candidate mobile node responds back with a new code. The code received from the candidate mobile is combined with the code that was originally sent to that mobile. The combined code is sent to the BS which modifies it with a mathematical expression and relays to the candidate mobile. The candidate mobile who is aware of this mathematical expression deciphers the code that it had sent/received to/from the mobile that has initiated the pairing. When it is successful, the channel is identified and the traffic path between the two nodes is established.

The random selection of two pairs may not alleviate the performance of the network and optimize the system capacity. This is because of the fact that channel conditions between the mobile node and the BS can vary to quite some extent. This may lead to disproportionate SINRs between the nodes and the BS, entirely depending on the channel quality. Hence User Grouping based on proportionate parameters is done by the network based on some Cost Function so as to optimize the performance.

3.5 D2D Using Fractional Frequency Reuse (FFR)

Interference is a significant issue in D2D communications. The interference is caused by devices in D2D mode with that in the normal cellular mode and the eNodeBs. It also depends a lot on the position of the D2D device in the cell, whether it is towards the exterior or towards the interior.

In [21], the radio resource allocation scheme using FFR is proposed. Different resources are allocated to the D2D UEs according to their location in the cell. If the D2D UE resides in the inner region of a cell, then they can use the frequency band that is not used by the eNodeB for relaying to the D2D UEs [24]. The ratio of different frequency reuse factor and the corresponding power level are optimized or adjusted adaptively according to the traffic load and user distribution. D2D and eNodeB relaying UEs located in the same cell do not interfere because radio resources from another frequency band are orthogonally allocated to the D2D and the eNodeBs relaying to the UEs.

3.6 D2D Using Cognitive Radio

In reference [22], there is an overview on how D2D communication can be realized over secondary users. With this concept, the primary users can only transceive via the BS. The secondary user can avail both D2D plus the BS transmission mode.

In reference [23] a joint subcarrier and power allocation method CR-D2D-MC for cognitive multicast with D2D communication coexisting with cellular networks have been proposed. The impact of imperfect spectrum sensing is considered in the proposed problem, which results in the capacity decrease of the cognitive multicast. The simulation results show that the proposed algorithm improves the spectrum efficiency and maintains a better trade-off between capacity and fairness for cognitive networks in a low algorithm complexity. Therefore, employing cognitive multicast based on D2D is able to explore more potential spectrum resources adequately to improve the system performance, and make it possible to satisfy the requirements of multiple kinds of high rate transmission.

3.7 Introducing SMNAT

SMNAT as a concept has been introduced in references [8, 12], which offers a global solution for various categories of cellular network. In SMNAT, the PHY layer is redesigned to take part directly in addressing and mobility management [13]. Similarly, in reference [10], it has been discussed how to implement SMNAT for a mobile network which serves M2M and H2H (Human-to-Human) users. In reference [11], SMNAT has been proposed for Vehicular communication network where D2D communication can be achieved at a group level. In reference [9], implementation scenario for a nano mobile network was brought in. In references [16–18] it has been demonstrated how an intelligent PHY layer can be conceived by using colour synthesis and conveyed electronically for use in mobile or fixed networks. In reference [19] a system model was developed showing how to use the smart PHY layer realized by colour coding for the purpose of joint source channel coding.

Of late, the focus of the telecom researchers is to make the PHY layer more generic, intelligent and adaptable for higher bandwidth. As an example, GFDM [25] implements a generalized Ortho Frequency Division Multiplexing (OFDM), which introduces additional degrees of freedom when choosing the system parameters. A technique called tail biting is employed to eliminate the need for additional guard periods that would be necessary in a conventional system, in order to compensate for filtering tails and prevent overlapping of subsequent symbols.

The purpose of SMNAT is to realize a smart and flexible PHY layer which can directly take part in some activities which are performed by layer 7 in the state-of-the-art networks. Take the example of mobility management itself which entails activities like device location tracking across cells, handover and channel management. All these activities demand some processing power required to actuate associated service logic at layer 7.

SMNAT aims to simulate layer 5 and 7 processes related to addressing and identification of a mobile node (UE) in the RAN, at the PHY layer [8, 13]. A mobile node is identified by a symbol located at a fixed phase and amplitude in the complex plane, a time slot in the time frame and a physical channel which are determined during provisioning of the UE. Figure 3.3 shows one Time Division Multiple Access (TDMA) frame, which includes T_{n+1} time slots (numbered from T_0 to T_n). The data rate required for the traffic burst is denoted by $f_d f_p$ denotes the sampling rates of the time slots before it is used to convey the traffic data. f_r denotes the refresh frequency for the given frame.

$$f_d >> f_p >> f_r$$

The modulation scheme implemented is a blend of M1 PSK (for outer ring) and M2 PSK (for inner ring), where M1<M2. Figure 3.3 shows the proposed constellation in the case of M1=8 and M2=4. The constellation diagram in Figure 3.4 pertains to a specific time slot, T1, of the frame. The frame corresponds to a specific subcarrier. Symbols of the outer ring are conveyed at a rate of f_r and the symbols of the inner ring are exchanged at a much higher rate, f_d.

In particular, one could set: $f_d = 8 \times f_r$. The outer ring comprises M1 symbols for user traffic, and the inner ring comprises M2 that will be used for the purpose of addressing the users. In the proposed multiple access scheme, a user is identified in the network with respect to the symbol coordinate in the

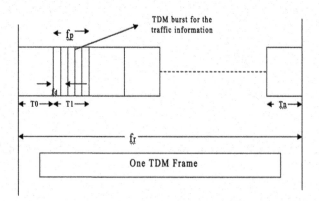

Figure 3.3   TDMA frame.

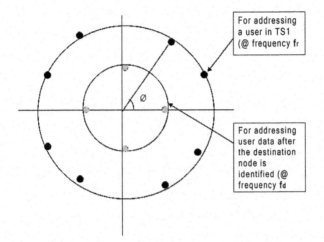

For addressing
a user in TS1
(@ frequency fr

For addressing
user data after
the destination
node is
identified (@
frequency fd

Figure 3.4 Constellation diagram.

complex plane. The 8PSK symbols are continuously rotated with 3/8 radians per symbol. The rotated symbols are defined in Equation 3.1.

$$\hat{s}_i = s_i \cdot e^{ji3\pi/8} \tag{3.1}$$

The modulated RF carrier is therefore given in Equation 3.2:

$$x(t') = \sqrt{\frac{2E_s}{T}} \, \mathrm{Re}[y(t').e^{j(2\pi f_0 t' + \varphi_0)}] \tag{3.2}$$

Where, E_s is the energy per modulating symbol, f_0 is the centre frequency and ϕ_0 is a random phase and is constant during one burst.

A single time slot can carry M1 symbols, used for addressing M1 users and later, the same time slot will be used to convey the symbols for data traffic employing M2 symbols of the inner ring of the complex plane after the mobile node is identified in the network.

Therefore, M1 users, which have been allocated M1 different symbols of the outer ring could use the same time slot and frequency sub-bands.

3.8 Network Architecture and the Processes

The network consists of access points which are Wireless transceivers/ repeaters supporting the Smart modulation scheme. As shown in Figure 3.5, multiple access points are interfaced to the local multiplexers. For the uplink,

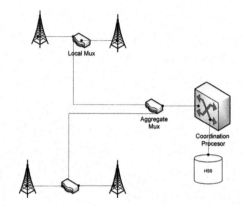

Figure 3.5 The multiplexer unit in the access network.

the local multiplexers aggregate the symbols in the time slots of the time frame pertaining to the different Absolute Radio Frequency Channel Number (AFCRNs). For the downlink, it acts as a transit node between the Coordination Processor and the Mobile Station (MS) to convey the broadcast signal related to all the AFCRNs/time frames to the MSs served by the network. Multiple Local Multiplexers converge to an Aggregate Multiplexer which transceives the time frames on both directions of the Coordination Processor [8] in the uplink and downlink. The high level architecture and the internal functional blocks of the Coordination Processor as in Figure 3.6 are summarized as below.

1. It receives the timeframe in the uplink from the Aggregate Multiplexer and analyses the PHY layer parameters for processing the response message and commence the termination leg of the call or SMS.
2. It formulates the aggregate frame for the downlink communication and relays it to the Aggregate Multiplexer.
3. It interfaces with the HLR/HSS and the VAS network elements to actuate user related tasks.
4. It takes part in the layer 7 signalling to enable authentication, support supplementary services and roaming.
5. It is responsible for CDR (Call Data Record) generation and interacts with the Mediation and Billing Systems.
6. It does dialled number analysis and actuates a TDM/VOIP breakout when needed.
7. It can actuate the functionality of a gateway router to receive calls from a state-of-the-art network. This is alike the functionality of Gateway MSC of 2G/3G networks.

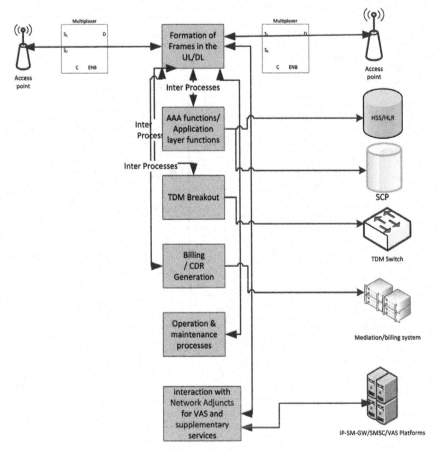

Figure 3.6 High level functionality of Coordination Processor.

3.8.1 Frame Formation in the Uplink

The control signalling messages are invoked by the mobile device only when it initiates a network operation, like Mobile origination Calls, SMS, etc. But no messages are generated for location update unlike the state of the art. The device never gets attached to the network. The network does not keep a track of the location and presence of the device continuously; hence one may eliminate the need of the continuous exchange of mobility management messages between the network and the device.

Figure 3.7 Illustrates how the frame is formed in the uplink for the addressing part. Multiple UEs (mobile devices) as in this diagram invoke

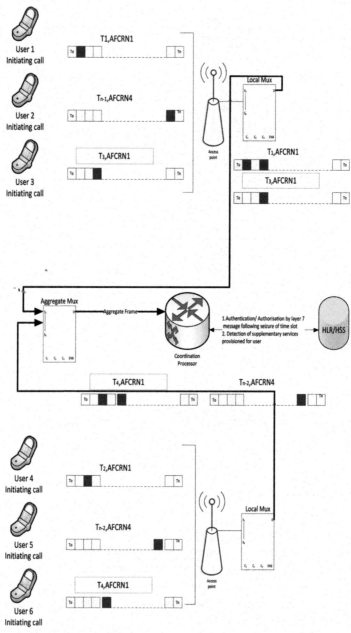

Figure 3.7   Aggregate frame formation in the uplink and user (A number) identification by Coordination Processor.

call requests. The UE first hunts for an available AFCRN where the time slot in the allocated position (as assigned to the user during the provisioning process) is identified. The symbol is then populated in the time slot and the UE generates a time frame by injecting only the specific symbol in that time slot in the time frame. Other time slots in the time frame are empty (as in Figure 3.7). The time frame pertaining to the available AFCRN is marked in black in Figure 3.7 which implies that the user equipment has successfully populated the assigned symbol (assigned to itself) in the time slot predefined for the user. The access point transfers the information to the local Multiplexer which aggregates the symbols ingressing from the user equipments and formulates a single time frame for each AFCRN. The Local Multiplexer formulates a time frame by injecting the symbols from the different users in the appropriate slots as per the time frames arriving from the user equipment. This time frame is transported to the Aggregate Multiplexer. The Aggregate Multiplexer synthesizes an aggregate time frame with respect to each AFCRN and forwards it to the Coordination Processor for analysis and call processing.

At this stage, it is impossible to identify the address of A party address (calling number of a particular user) from the symbol coordinate and timeslot, because the AFCRN is variable. Hence the Coordination Processor cannot commence authentication. It first needs to secure the time slot in both uplink/downlink directions between the MS and itself. Hence it generates a response message in downlink towards the MS using the same AFCRN number that was used by the A party in the during the uplink layer 1 message for call initiation. However, the particular AFCRN may not be available because it may already be in use as traffic channel by other users. In this case, the Coordination Processor does not generate the response message towards the MS. The MS waits for the timer expiry (in milliseconds) and restarts the scan process and re-initiates a new message with another available AFCRN. In this process, it skips the AFCRN that was previously determined as unavailable. Following a successful layer 1 handshake between UE and the Coordination Processor, the time slot pertaining to the AFCRN is used further as traffic channel. The UE generates a layer 7 message towards the Coordination Processor encapsulated in the time slot that was allocated following the layer 1 handshake process as described earlier. The data conveyance materializes following M2 PSK modulation on the inner ring of the constellation diagram (Figure 3.3). UE populates A party addresses (calling address in E.212 format), authentication parameters, dialled numbers, supplementary service information in this message, so that the Coordination

Processor can liaise with the HSS/HLR to actuate authentication, authorization for A party. Subsequently, it analyses the B Party (dialled) number and figures out the outgoing channel by carrying out dialled digit analysis. The above process deviates from mobile originating call procedure followed by the state-of-the-art Mobile Networks. Generally in conventional networks, the serving MSC/MME/CSCF generates an ISUP/SIP message directly for call termination towards the B party address. There is no interrogation towards the HLR//HSS. Only in case of Home routing scenarios, the call is routed towards the Home network where the SCP (Service Control point) is queried to generate a "Connect To" number.

3.8.2 Frame Formation in the Downlink

The downlink process of how the call matures at the downlink towards B Party (called address) UE is shown in Figure 3.8. When the call arrives at the Coordination Processor, it first analyses if the B party address is served by the same mobile network, i.e., the smart network. If yes, then it interrogates the HSS/HLR for obtaining the symbol coordinate, time slot and the primary AFCRN allocated for the B Party. The supplementary and the tele-services assigned for the B party are also downloaded by the HSS/HLR towards the Coordination Processor. Subsequently, the Coordination Processor injects the symbol of B party in time slot (preconfigured for B number) in the time frame for the particular AFCRN in the downlink. In case the primary AFCRN is busy, then alternative AFCRNs are tried according to the predefined frequency scanning schemes. Note that this time frame also carries the symbols from other users in the respective time slots. Aggregate frame is formed and is

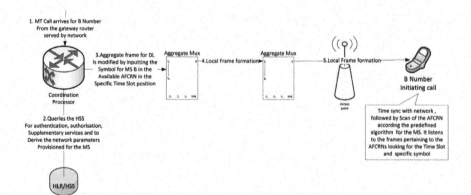

Figure 3.8   Frame formation in the downlink

routed to the Aggregate Multiplexer. The Aggregate Multiplexer broadcasts the time frames pertaining to all the AFCRNs in the Downlink via the Local Multiplexer and access points. B party scans the AFCRNs and looks out for the specific symbol contained in the specific time slot assigned for B number. If it finds one, then it formulates a response message by injecting the same symbol in the same time slot position in the UL. If the time slot (pertaining to the AFCRN) is busy, there is no response back from B party. The Coordination Processor has a time out and tries to page B party again with an alternate AFCRN according to the frequency scanning logic. As AFCRN is not fixed, one cannot ascertain at this stage seeing the radio layer PHY layer message whether the page message is intended for the particular B party address or some other users. The page message can indeed be meant for another user provisioned with the same symbol coordinate and time slot position. Hence, multiple users can simultaneously try to generate the response message. But once the Coordination Processor receives the layer 1 page response, the time slot related to the AFCRN will be reserved for use by the Coordination Processor. The time slot/AFCRN is also reserved at the UE end. The subsequent layer 7 message generated by the Coordination Processor will bear the B party address (E.212) as well as the A party address (E.164 for the CLI) and the supplementary service information. The specific UE (the original B party) will respond to this layer 7 message. The other UEs which reserved the time slot will release it when it ascertains from the layer 7 message carried by the time slot that the page was intended for another user (i.e., the B party). This completes the addressing process in the downlink.

3.8.3 Integration of SMNAT with 5G Cores

It is imperative that the success behind any new technology lies in its interoperability and integration possibility with the existing or evolving networks. As a reference, in [14], the primary challenges for interworking between LTE core network and legacy core network are discussed. SMNAT is oriented on the PHY layer to implement the processes of addressing, mobility management and data exchange. On the contrary, 3GPPs vision on mobility management on 5G is founded on the application layer, more specifically the NAS (Non Access Stratum). The NAS defines the basic processes for mobility management for EPC (Enhanced Packet Core) between the UE and the MME. The two topologies are ideologically converse. Due to this reason it is recommended to follow a two stepped approach for integration of the two technologies.

Short-term approach

WLAN can be integrated by the EPC core as it permits integration of non 3GPP untrusted network via the ePDG (evolved Packet Data Gateway). In Figure 3.9, the end-to-end integration between SMNAT and EPC is shown.

The UE establishes a communication with HSS through ePDG for actuating EPC authentication. At SMNAT end, RADIUS based authentication is followed alike the EAP-AKA mechanism. The conversion rules from RADIUS to DIAMETER (EPC) follow the GSM IR 61 recommendations.

The data transfer between the UE and the IMS core is established via the S2b interface between the ePDG and the PGW.

The call flow is shown in Figure 3.10. It can be seen that two processes are covered.

1. Authentication
2. Data transfer

It may be interesting to note that mobility management messages are not exchanged with EPC. In a state-of-the-art (LTE) network, the UE actuate NAS signalling with the EPC. NAS, which is a layer 7 process, is not implemented in SMNAT. Rather SMNAT is dependent on a process where location management is realized via layer 1. This is the reason, the mobility management messages like Location Update/Location Cancellation are not seen (Figure 3.10). SMNAT can directly fit in the architecture which has been realized for integrating Wi-Fi with EPC.

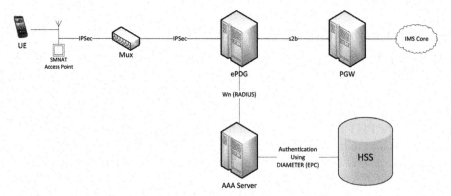

Figure 3.9   Integration of SMNAT with EPC.

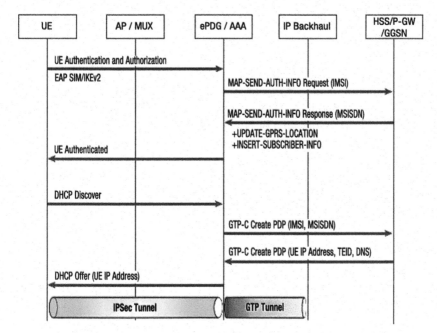

Figure 3.10   Call flow between UE, SMNAT access and EPC core.

Long-term approach

The long-term approach is to evolve the eNodeBs and MMEs as to be compatible with SMNAT architecture. As the mobility management part is simplified, these smart eNodeBs will thrive on PHY layer addressing, rather than the communication on NAS. The MMEs will be closer to the definition done for Coordination Processor, which can directly engage itself in PHY layer addressing and mobility management. Hence eNodeBs/Home eNodeB and MMEs will be lighter in terms of power consumption and processor capacity than the state-of-the-art networks.

3.9 Implementation of SMNAT for In-Band–D2D and Interoperability with WISDOM

This section brings out how SMNAT can be used to realize D2D communication, which can solve some issues and achieve some specific objects which the state-of-the-art D2D technologies cannot attain. Contrary to the work done so far on D2D where the peering is only possible between the two devices in physical proximity, SMNAT offers a more versatile solution where the D2D leg also could be established between two UEs which are not in vicinity. The following section discusses how this can be achieved.

SMNAT can be deployed as an access and core network layer (Figure 3.11), located on the same network plane as the WISDOM 5G [26–28] based on cognitive communication. In such a scenario, SMNAT will be dedicated for D2D communication, while WISDOM will be responsible for global cellular roaming within the network area or beyond.

The available spectrum will be allocated between SMNAT and WISDOM. The ratio of the spectrum allocation can be determined through the process of network planning and can vary according to the traffic characteristics and the business requirement of the particular operator.

To decide whether D2D leg will be established or not, there are two possible options.

1. Decision by the user
2. Decision by the network based on cooperative communication

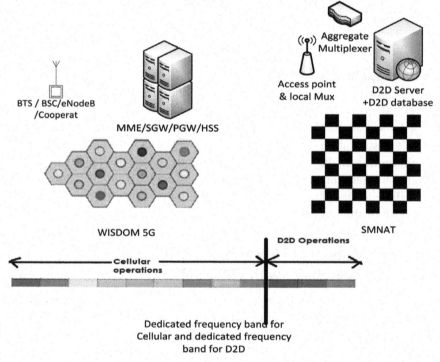

Figure 3.11   Coexistence of WISDOM 5G and SMNAT.

When any device initiates a communication process on data, it will have two options. It will be given a chance to try a D2D communication with the incentive of a faster communication process or lesser data charges (if offered by the network operator).

If the device does not opt for the D2D option, then the session will be established via the normal process of EPC signalling. Subsequently, the EPC can further decide whether to terminate the session to the candidate device via LTE/LTE-A, or the network will send instructions to the device rather to automatically initiate a D2D session.

3.10 Description of network elements of SMNAT and the Call Flow for Session Establishment

SMNAT [12] [8] has a unique design of the cellular network which represents a checkerboard. There are access points in each cell, which via the Frame Aggregator can communicate with the D2D server. The D2D server is

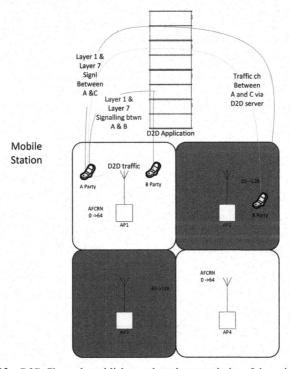

Figure 3.12   D2D Channel establishment based on proximity of the paired device.

interfaced to a D2D database which holds the mapping between the E.164 address of the mobile device and the SMNAT coordinates namely the symbol coordinate, time slot and AFCRN.

As evident from Figure 3.12, D2D transport channel can be established directly between two nodes in case they are in proximity (same or different cell), but facilitated by the SMNAT Access and core network (for signalling). In case the devices are not in the direct signal range of each other, then the traffic channel will be established via the SMNAT Access and core (access point Frame Multiplexer D2D Server). But this process uses the addressing and mobility management methods pertaining to the PHY layer as defined by SMNAT.

In Figure 3.13, the end-to-end call flow for D2D channel establishment is shown, which is self-explanatory. D2D server is a core network application platform which is responsible in establishing the pair, either directly between

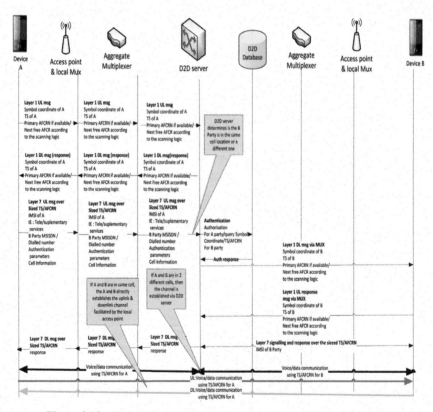

Figure 3.13   End–to-End communication channel establishment for D2D.

the two devices or via the core network using PHY layer addressing and mobility management.

3.11 Decision by the Network to Initiate D2D Based on Cooperative Communications

In case the user attempts (by default) to establish communication through BS via the WISDOM 5G [26–28] network, then the 5G Network will first check if the session can be terminated by the D2D channel. This implies that the network will check if it can offload from cellular to D2D for capacity and optimization reasons. This can be done with the help of Diameter interaction between the HSS and the D2D server. A high level process is shown in Figure 3.14.

3.12 Security Aspects in Light of Cooperative Communication between 2 Devices in Cellular D2D Mode

As explained before, D2D communication can be established in the direct mode or via the network depending upon the proximity of the devices,

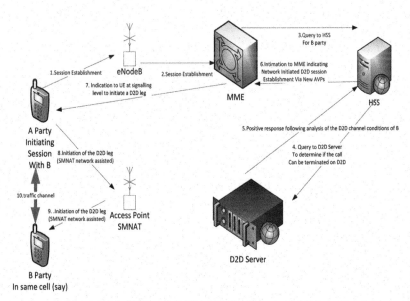

Figure 3.14　Network initiated D2D channel establishment.

channel conditions and location (both devices in the same cell or different cells). Security has become a primary concern especially for the D2D communication.

In case of the network assisted D2D communication, the network can enforce the security methods that are in place for cellular communication. The integration of SMNAT with the EPC has been shown in Figure 3.9, where the conventional EAP AKA method meant for traditional WLAN network can be used. The ePDG (Evolved Packet Data Gateway)/AAA platform modifies the RADIUS based authentication messages to AKA based EPC authentication parameters over DIAMETER directly with the HSS. The EPC authentication vectors including ciphering keys that are downloaded from the HSS will be retranslated and compared for the purpose of authenticating the UE.

However, for the direct mode of D2D communication, where two devices need to pair with each other using the cellular channel, the procedures become intricate. Generation of symmetric keys for the communicating peers without key exchange is challenging. One of the methods to generate the keys is based on random variations of wireless channel like Channel State Information (CSI). Most of these methods usually use the measurement results of individual subcarriers. But this is not robust, given the fact that the subcarriers in proximity can have similar channel conditions and have strong correlation dependence. Due to this fact there can be repeated segments in the key, which may be easy to crack.

Another approach is to use RSS key generation mechanism based on channel measurements. But the main issue is that it is not designed to meet the requirements of a system demanding key generation at massive scale. Bit generation rate is low as a single sample can provide a single RSS value. This also becomes vulnerable if the UE is static and the channel conditions are predictable and the channel variations are predictable. Hence the endeavour of the researchers is to realize a method of key generation based on CSI, but not based on a single subcarrier. In reference [29], a new algorithm termed as KEEP has been proposed. KEEP aims to actuate the key generation process by incorporating the following sub-processes.

1. Dropping the inconsistent bits by exploiting the correlation of CSI measurements from multiple carriers. This is executed by a federated filtration method.

2. Implementation of Universal has key for validating consistency of bit streams between two communicating channels. A pair of devices

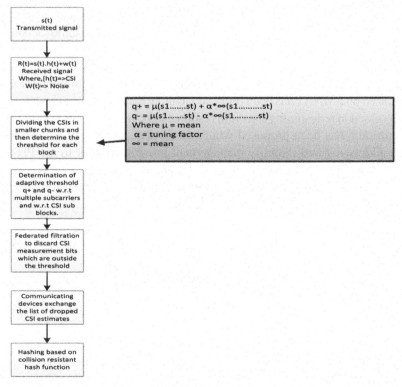

Figure 3.15 Key generation process.

exchanges only a part of the hashed keys, which makes it difficult for the attacker to guess and formulate the whole key.

3. Using methods of key recombination and adaptive quantization to generate secret bit from a large number of subcarriers. The bit mismatch rate with the process is low. Moreover, it eliminates the correlation of bit streams from multiple subcarriers and reduces predictivity.

The key generation process based adaptive quantization procedure, CSI sub-blocking and hashing is captured in the following algorithm in Figure 3.15.

3.13 Summary

At present, mobility management is explicitly in the domain of the application layer. This chapter introduces SMNAT where it is proposed to shift from this paradigm and knit this functionality with the PHY layer. This entails simplification of network architecture and associated processes and helps in attaining a network which has a degree of location and presence agnosticism. The interaction between the device and the network occurs only when one of them initiates a network operation for a call, SMS or other services. The handover operation is also simplified. Further, a novel D2D communication concept based on SMNAT has been proposed. Unlike the existing researches in D2D, engrossed in redesigning the radio resource allocation process, a parallel network stratum has been proposed which can actuate D2D communication based on addressing and mobility management processes closer to the PHY layer. The processing cycles in the network will diminish, which creates extra room to cater to new customers. This is aligned with our endeavour to realize a network which is cleaner, greener and leaner.

References

[1] 3GPP 23.228 Architecture and main flows for a IMS system. Rel 11, version 11.5.0, 2012-06-22.

[2] Rich Communication Suite 5.2 Advanced Communications Services and Client Specification, GSMA.

[3] Amit Kumar; Yunfei Liu; Jyotsna Sengupta; Divya; Evolution of Mobile Wireless Communication Networks:1G to 4G, IJECT Vol. 1, Issue 1, December 2010, pp. 68–72.

[4] 3GPP TS 24.301: Non-Access-Stratum (NAS) protocol for Evolved Packet System (EPS); Stage 3. Rel 11, version 11.3.0, 2012-06-27.

[5] 3GPP TR 22.803 V12.2.0 (2013-06), Feasibility Study for Proximity Services (ProSe), (Release 12).

[6] 3GPP TR 22.888 V12.0.0 (2013-03), Study on enhancements for Machine-Type Communications (MTC) (Release 12).

[7] Rajarshi Sanyal and Ramjee Prasad, Enabling Cellular Device-to-Device data exchange on WISDOM 5G by actuating cooperative communication based on SMNAT, paper submitted.

[8] Sanyal, R., Cianca, E. and Prasad, 'Smart mobile networks with intelligent physical layer actuating mobility management', Int. J. Mobile Network Design and Innovation, Inderscience, 2012 Vol.4, No.2, pp. 91–108.

[9] Sanyal, R., Cianca, E. and Prasad, R 'Nano mobile network based on Smart Location Management and Addressing', Int. J. Mobile Network Design and Innovation, Inderscience, Vol. 4, No. 3, 2012.

[10] Sanyal, R., Cianca, E. and Prasad, R, 'Challenges for Coexistence of Machine-to-Machine and Human-to-Human Applications in Mobile Network: Concept of Smart mobility management, International Journal of Distributed Sensor Networks, Hindawi publishers, Volume 2012 (2012), Article ID 830371, 12 pagesdoi:10.1155/2012/830371.

[11] Sanyal, R., Cianca, E. and Prasad, 'Closed User Group Automotive communication network based on addressing at physical layer', International Journal of Interdisciplinary Telecommunications and Networking. IGI Global publishers USA, Volume 4, Issue 4, 2012.

[12] R. Sanyal, E. Cianca, and R. Prasad, "Beyond LTE: next generation multiple access technology with intelligent lower layers," in Proceedings of ICWN 2011, the World Congress Computer Science, Computer Engineering and Applied Computing and International Conference, 2011, p. 474.

[13] Rajarshi Sanyal, Ernestina Cianca, Ramjee Prasad "Rendering Intelligence at Physical Layer for Smart presented at IARIA conference on wireless networks 2011, France, published at IEEE digital library.

[14] Rajarshi Sanyal, 'Challenges in Interoperability and Roaming between LTE - Legacy core for mobility management, Routing, Real Time Charging' ITU WORLD 2011, Geneva, 26th October 2011, published in IEEE Explore.

[15] Sanyal, R., Cianca, E. and Prasad, 'Novel WLL architecture based on Colour Pixel Multiple Access implemented on a Terrestrial Video Network as the overlay', International Journal of Interdisciplinary Telecommunications and Networking, IGI Global publishers USA Volume 5, Issue 1, 2013.

[16] Rajarshi Sanyal, Ernestina Cianca, Ramjee Prasad, Colour Pixel Multiple Access- A multiple Access Technology for Next Generation Mobile Networks', International Conference on Wireless Networks (ICWN 09), 2009 World Congress Computer Science, Computer Engineering and Applied Computing at USA.

[17] Rajarshi Sanyal, 'Framework for Realizing mobile and computer communication through color signals', SAM 2006, World Congress in Computer Science, Computer Engineering and Applied Computing, USA.

[18] Rajarshi Sanyal, Mobile and Computer Communications Through Colour Signals - An Approach Note. WINSYS 2006: 73–78. Portugal.

[19] Rajarshi Sanyal, Transmission of Audio signals from multiple sources riding on single lossless streaming video channel using Phase Vocoder by

applying Joint Source Channel Coding technique, (CIC07): 2007 World Congress in Computer Science, Computer Engineering and Applied Computing in Las Vegas USA.

[20] Afif OSSEIRAN, Klaus DOPPLER, Cassio RIBEIRO2, Ming XIAO, Mikael SKOGLUND, Jawad MANSSOUR, Advances in Device-to-Device Communications and Network Coding for IMT-Advanced, ICT-MobileSummit 2009 Conference Proceedings.

[21] Hyang Sin Chae, Jaheon Gu, Bum-Gon Choi, and Min Young Chung, Radio Resource Allocation Scheme for Device-to-Device Communication in Cellular Networks Using Fractional Frequency Reuse, 2011 17th Asia-Pacific Conference on Communications (APCC), 2nd–5th October 2011.

[22] Peng Cheng; Dept. of Electron. Eng., Shanghai Jiao Tong Univ., Shanghai, China; Lei Deng; Hui Yu; Youyun Xu Resource allocation for cognitive networks with D2D communication: An evolutionary approach, Wireless Communications and Networking Conference (WCNC), 2012 IEEE.

[23] Yueyun Chen1, Xiangyun Xu1, Qun Lei1, Joint Subcarriers and Power Allocation with Imperfect Spectrum Sensing for Cognitive D2D Wireless Multicast, KSII TRANSACTIONS ON INTERNET AND INFORMATION SYSTEMS VOL. 7, NO. 7, Jul. 2013.

[24] Tarun Bansal, Karthikeyan Sundaresan, Sampath Rangarajan, Prasun Sinha, R2D2: Embracing Device-to-Device Communication in Next Generation Cellular Networks, Proc. of IEEE INFOCOM, Toronto, Canada, Apr 2014.

[25] G. Fettweis, M. Krondorf, and S. Bittner, "GFDM – Generalized Frequency Division Multiplexing," in Proc. IEEE 69th Vehicular Technology Conf. VTC Spring 2009.

[26] Cornelia-Ionela Badoi, Neeli Prasad, Victor Croitoru, Ramjee Prasad, 5G Based on Cognitive Radio, Wireless Pers Commun (2011) 57: 441–464.

[27] Ramjee Prasad, Global ICT Standardisation Forum for India (GISFI) and 5G Standardization, Journal of ICT Standardization, River Publishers, Vol: 1 Issue: 2, June 2013.

[28] Ramjee Prasad and Albena Mihovska, Challenges to 5G standardization. ITU News, Issue N° 10 December 2013.

[29] W. Xi, X. Li, C. Qian, J. Han KEEP: Fast Secret Key Extraction Protocol for D2D Communication, S. Tang, K. Zhao, Proceedings of IEEE/ACM International Symposium on Quality of Service (IWQoS), 2014.

4

Dynamic Spectrum Management
and mm-WAVES

Radio spectrum is a prime factor in driving the growth of mobile services. The success of 5G network is based on the unconstrained availability of spectrum. About 1200 MHz of spectrum in the frequency bands below 5 GHz has been identified for IMT services during World Administrative Radio Conference (WARC)-92, World Radio communication Conference (WRC)-2000 and WRC-2007. These frequency bands are 450–470 MHz, 698–960 MHz, 1710–2025 MHz, 2110–2200 MHz, 2300–2400 MHz, 2500–2690 MHz, and 3400–3600 MHz. The identified spectrum is non-contiguous and scattered in different frequency bands from 450 MHz to 3.4 GHz. However, the actual allocation is ranging between the frequency band 700 MHz and 2.6 GHz. The irony is that these identified frequency bands have already been allocated to legacy services long back. Therefore, no vacant spectrum is available especially below 6 GHz at present for mobile communications. The options available to enhance the spectrum availability for 5G communications are spectrum re-farming, spectrum sharing and use of cognitive radio technology. Moreover, this identified 1200 MHz non-contiguous spectrum could not hold the pressure of high mobile data growth, demand for convergence of different varieties of services and speed as envisaged in 5G network. Assigning a new radio spectrum is crucial to meet the expected demands for future 5G networks. This is possible by exploiting higher microwave frequencies, referred as millimeter (mm-wave) bands. Therefore, mm-frequency band is the obvious and the most preferred band for 5G network.

The 5G network envisages as a combination of several micro, pico and femto cells embedded within a macro cell. According to physical law, coverage decreases with increasing frequency. The mm-waves can be divided into different categories, the first one ranging between 20 and 40 GHz frequency bands for micro sites and the other one is around 60 GHz frequency band for pico and femto cell sites.

With the increase in the number of wireless devices, the number of wireless connections and high data rate networks rises. This leads to the two important factors spectrum demand and spectrum congestion, turning out to be the two critical challenges for the forthcoming wireless communication world. Simultaneously the user's requirements such as high multimedia data rate transmission based on the bandwidth demanding applications will make the future wireless networks to suffer from the spectrum scarcity.

4.1 Command and Control Method

The conventional method for allocating spectrum is known as "Command and Control Method" shown in the Figure 4.1. There are some countries following this technique of spectrum allocation. In this method radio spectrum is divided into different spectrum bands that are allowed to specific radio communication services such as satellite services, mobile, broadcast on an exclusive basis.

This method guarantees that the radio frequency spectrum will be exclusively licensed to an authorized user and can use spectrum without any interference [1].

This method of spectrum allocation is not efficient because [1]:

- Spectrum assigned to a particular radio communication service cannot be replaced by other services even though it is witnessed that spectrum is underutilized.
- There is no possibility of questioning the user once the spectrum is allocated to him (during the licensing period) as per the norms, provided he fulfills the terms and conditions.

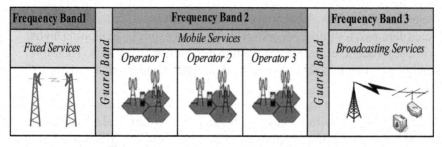

Figure 4.1 Command and Control Method [1].

- This method does not make spectrum to be utilized efficiently in rural areas as the spectrum utilization is heavy in the urban regions and underutilized in the rural areas.

It is sad to see the spectrum underutilized and not accessible to all. It is difficult for some nations to provide 4G services itself. There is a need for taking serious steps in handling spectrum issues by implementing sophisticated technologies for the development of nation. In these cases the techniques like spectrum trading would be a successful solution. This will only lead to the development of 5G communications in these types of countries.

4.2 Spectrum Sharing

The demand for multimedia content and information processing, services such as e-education and e-health, mobile broadcasting, enormous increase in the electronic gadgets necessitate efficient use of all available and usable frequency spectrum. The new generation of mobile broadband networks will require supporting higher data throughput rates.

Many sophisticated technologies have been implemented for making the efficient use of available spectrum. For example, line-of-sight (LOS) systems are usable up to 100 GHz now. Running down the size of electronic components and systems introduces the multiple frequency bands in single equipment leading to the efficient use of available spectrum by the enhanced dynamic sharing of frequency bands.

Spectrum management should be in such a way that there should be always optimum spectral sharing. Greater sharing of frequencies and bands allows more data to be sent by different users in the same amount of available spectrum.

Spectrum sharing has basically three dimensions: frequency, time and location. The Collective Use of Spectrum (CUS) allows spectrum to be used by more than one user simultaneously without requiring a license. Some of the examples that come under spectrum sharing are frequency reuse concept in the existing telecom networks, FDMA, TDMA. Another important challenge is the sharing of spectrum among the heterogeneous networks. While it is easier to achieve efficient and successful spectrum sharing among the homogeneous or similar networks or applications, there arises complexity in heterogeneous networks [2].

The spectrum sharing methods are classified into three categories based on based on the priority level of accessing the radio spectrum as follows [2]:

    a. Horizontal spectrum sharing: all the devices have equal rights to access the spectrum.

    b. Vertical spectrum handover only: the primary users are allotted priorities to access the spectrum.

    c. Hierarchical spectrum sharing: it is an enhanced variant of the vertical spectrum sharing.

4.2.1 Spectrum Using SDR and Cognitive Radio – Dynamic Sharing

Evolution of software defined radio (SDR) and cognitive radio (CR) are the two major milestones in the mobile communications. Dynamic sharing of spectrum improves the spectrum efficiency and the above mentioned technologies play a vital role in this aspect.

Conventionally, transmitters were tuned to specific frequencies, and facilities for multiple frequencies would cost high. But after the development of these technologies, tuning the transmitters to the multiple frequencies has become easier, i.e., switching to the different frequencies in a dynamic way would be possible at a reasonable cost.

Cognitive radio first detects the occupation of the channel, and if it is occupied, it helps the users to switch to the other vacant channels. Also the carrier signals are sensed regularly for usage in other. There is always a need of large amount of spectrum in case of emergency or public safety conditions compared to that of normal conditions. In these emergency cases, dynamic sharing of spectrum would be a promising solution. In some countries spectrum regulators are used for the encouraging dynamic sharing spectrum with public safety requirements. It is to be noted that CR is a combination of administrative (regulatory), technical, and market based techniques to enhance the efficiency of spectrum utilization [2].

Another area of utility for dynamic sharing is White Spaces (TV Band). Normally, the TV broadcasters repeat the same channel/carrier at relatively longer distances, to avoid any interference especially at the border/edger of the coverage areas that are on the border of two adjacent broadcast transmissions on same channel. However, there are very few receivers in this area, and the spectrum utility is not effective and could be utilized for other purposes.

The broadcasters are generally quite protective for their signal trans-missions, even in areas beyond the theoretical coverage areas. Hence, only low power systems that cause minimal interference can be considered for shared usage with the TV spectrum. However, gradually with time building collective confidence amongst the users that includes the broadcasters, higher power based systems could be considered [2].

4.3 Spectrum Trading

Spectrum trading is a case of spectrum sharing with the involvement of commercial activities. Spectrum trading is found to be a more economical way of efficient use of spectrum. It is an option through which flexibility can be increased and spectrum assigned to a particular service, and can be easily transferred for other usage. To explain it in brief, spectrum trading is a market based mechanism where buyers and the sellers determine the assignments of spectrum and its uses in which seller transfers the right of spectrum usage, in full or part, to buyer while retaining the ownership. In many countries spectrum trading is already running and the trading procedure is confined to specific bands, which are in demand for commercial use with specified conditions. Spectrum trading improves the efficiency and facilitates new services to enter in the market by making slight modification in the regulatory provisions [2].

The difference between spectrum sharing and spectrum trading can be explained as follows:

In spectrum trading the usage rights are transferred completely from the seller for a specified period. However, in spectrum sharing buyer gets a temporary right of spectrum usage with the exclusive rights resting with the seller. Trading becomes effective only when it is clubbed with liberalization. Spectrum trading can be implemented if there is solid base in understanding advanced technologies and operating systems as the spectrum flexibility demands new approaches and practical methods for monitoring compliance, enforcement and conflict resolution [2].

4.3.1 Spectrum Trading Merits

The merits of spectrum trading are as follows [2]:

- Improves efficient spectrum usage
- Facilitates the evaluation of spectrum licenses, and gaining knowledge of market value of spectrum

- Quicker process, with better and faster decision-making by those with information
- Removes barriers to entry by allowing small operators and start-ups to acquire spectrum rights of use more readily, thereby facilitating the development of market competition
- There is an opportunity for more rapid redeployment and faster access for spectrum
- Encourages new technologies to gain access to spectrum more quickly
- Existing operators gain an opportunity to sell unused or under-used spectrum and make more flexible use of spectrum
- Reduction in the transactions costs of acquiring rights to use spectrum
- Allows operators increased flexibility to accommodate shifting demand driven by market changes.

4.4 Cognitive Radio

IEEE approved definition of cognitive radio (CR) is a radio in which communication systems are aware of their environment and internal state, and can make decisions about their radio operation based on that information and predefined objectives. The environmental information may not include location information related to communication systems. Cognitive radio is a very good solution for increasing the spectrum utilization.

Cognitive radios should be able to self-organize their communication based on sensing and reconfigurable functions as stated below [3]:

- *Spectrum resource management:* this scheme is necessary to manage and organize efficiently spectrum holes information among cognitive radios.
- *Security management:* cognitive radio networks (CRNs) are heterogeneous networks in essence and this heterogeneous property introduces a lot of security issues. So this scheme helps in providing security functions in dynamic environment.
- *Mobility and connection management:* this scheme can help neighbourhood discovery, detect available Internet access, and support vertical handoffs, which help cognitive radios to select route and networks.

4.4.1 CR Device Concept

This section explains the features of CR whose implementation in a single device offers a very smart and high performance user terminal – CR terminal. The Figure 4.2 shows the CR properties.

A. Spectrum sensing

Spectrum sensing operation can be divided into the three step functions [4]:

- *Signal Detection:* In this step of operation existence of the signal is sensed. There is no need to know the type of signal in this step.
- *Signal Classification:* In this step of operation the type of signal is detected, which is done by extracting the features of the signal.

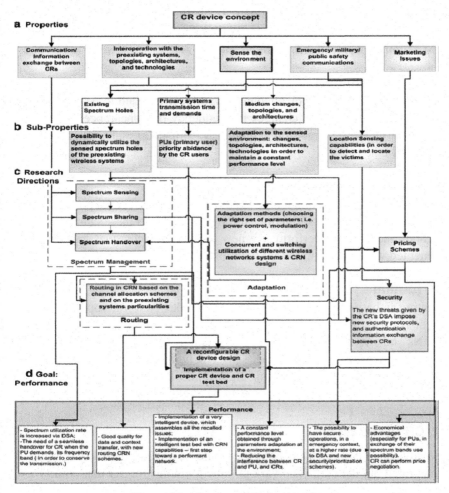

Figure 4.2   Concept of CR Device [4].

- *Channel availability decision:* In this channel availability is detected. Once the free channels are detected, the step next to it is, sharing the spectrum holes which can be achieved by the spectrum allocation scheme.

The CR technology also brings new *security and pricing challenges* which are shown in the Figure 4.2.

- New security threats appear with the dynamic spectrum access concept, as well as the CR's authentication needs.
- The pricing is very much influenced by the used channel allocation scheme. Additionally, CRs must be designed with strong capabilities to negotiate the available channels' price.

B. Spectrum handover
The phenomenon of frequency changing dynamically is said to be spectrum handover. A secondary user changes its frequency on appearance of a primary user or due to transmission degradation. This necessitates designing a handover scheme [4].

C. Environment adaptation
Different changes like topological changes, noise or interference power may occur while sensing the information. In order to adapt to these changes and to maintain the constant performance new adaptation techniques have to be implemented which is an important point of concern [4].

D. CR routing
CR routing is based on the requirement for CR device to interoperate with different systems, and is influenced by the spectrum sharing techniques [4].

CRNs inherit the PSs (Primary Systems) network characteristics: infrastructure - based, mesh, ad-hoc, sensor networks, etc. and these architecture types impose a specific routing algorithm, which must also include the CR devices and the possibility for a CR to be a relay node for another CR.

4.4.2 CR based on 5G

As already stated earlier CR technology would be a major modality to build the integrated 5G network. The various functionalities for 5G that could be met with CR usage are as follows [4]:

- Advanced PHY and MAC technologies.
- Implementation of novel and flexible protocols.

- Capacity to support homogenous and heterogeneous systems.
- Adaptation to different changes like environment changes, dynamic frequency changes, etc.

Correlation between WISDOM and CR in reference to 5G could be given as:

"5G brings the convergence concept through WISDOM and CR represents the technological tool to implement it." The 5G technology eliminates the radio terminals that are specific to particular wireless technologies and proposes a universal terminal which must include all of the predecessor features in a single device. This terminal convergence is supported by the users' needs and demands and is strongly found in CR terminal [4].

There are many issues that still remain to be addressed [4]:

- How to connect the CR terminal to the wired networks?
- How to reach the maximum 5G's 1 Tera bps data rate threshold when using the CR technology at the access level?
- How to implement the good techniques in order to combine the flows coming from multiple access networks?

4.5 Millimetre Waves

Most of the radio communications including TV, satellite communications GPS, Bluetooth are utilizing frequency band ranging from 300 MHz to 3 GHz. But this band is getting crowded and the focus is on releasing and utilizing the additional spectrum. In mm-waves are the promising solution for this problem.

The spectrum bands identified under the IMT umbrella do not have the capacity to carry such enormous data required for 5G services. Therefore, mm-waves could be the candidate bands for 5G mobile communications due to high data carrying capacity. The mm-waves have the following advantages [5, 6]:

(a) Not much operation at mm-waves so more spectrum is available at mm-waves
(b) Very large blocks of contiguous spectrum to support future applications.
(c) Due to high attenuation in free space, frequency reuse is possible at shorter distance
(d) Spatial resolution is better at mm-waves hardware with CMOS technology

(e) Advancement in semiconductor technology allows low cost equipment
(f) Small wavelength makes possible use of large antenna arrays for adaptive beam forming
(g) Small size of antenna at mm-waves facilitates easy integration on chip and installation at suitable locations.

In mm-waves allow larger bandwidth and offer high data transfer and low latency rate that are suitable for high speed reliable Internet services. The small wavelength facilitates small size antenna and other part of radio hardware, which reduces costs and also easy to install. The transmitter's antenna would be like a lamppost, which could be installed on building, street lamppost, etc. [6].

High directionality attained in this band can be used to increase spatial multiplexing. The size of antenna required for a mm-waves radio can be one-tenth or less of an equivalent lower frequency radio which is an advantage to the manufactures to build smaller and lighter systems.

Beam width is the measure of how a transmitted beam spreads out as it gets farther from its point of origin. But due to limited availability of radio frequency (RF) bands the fifth generation wireless communication systems will move to ultra-high capacity mm-wave bands. High frequency makes mm-wave band more attractive for wireless communication system and these frequencies are used in terrestrial and satellite communications. Wireless products that use millimeter waves already exist for fixed, LOS transmissions

But the absorption rate of the mm-wave electromagnetic signal poses great challenges for their utilization in the non-LOS and mobile connections. On the other hand, high directionality achieved in this band can be used to increase spatial multiplexing. Wireless backhaul will be another key enabler of 5G-mm-wave small cells [8].

Within the mm frequencies, the frequency band of 60 GHz has attracted the researchers to work with, as the large amounts of bandwidth are unallocated in this band, bandwidths that are required for communication systems at the intended data rates of 100 Mbps and above. Also, another advantage of 60-GHz band is due to a physical property of the propagation channel at this frequency that provides a natural way for reduction of frequency reuse factor, which tends to compact cell size [8].

It is a general property of the mm-wave propagation that the behaviour of the propagation rays is well characterized by the geometric optics. That is, the waves do not penetrate the walls or other obstacles and wave reflection is the main mechanism leading to a multipath [8]. In mm-waves have the potential to support broad-band service access which is especially relevant because of the advent of Broadband Integrated Service Digital Network (B-ISDN).

With the development of personal wireless communication systems, two things are appearing to be significant:

- Exploiting high frequency bands, such as mm-waves to provide broadband for high rate data transmission.
- To integrate multi-tasks in one system which greatly extend the application of wireless device.

The utility of mm-waves for the micro cells that form the WISDOM based GIMCV are well positioned to be served by these mm-waves. It has been elaborated in these following points:

- It is relatively easy to get licenses for big blocks of mm-wave spectrum, which would allow carriers to deploy large backhaul pipes over 1 Gbps in size. While a single small cell may not need that much capacity, the complexity of heterogeneous networks will require daisy-chaining many small cells together, each cell passing its load down the line.
- Small cell backhaul makes the best use high frequency characteristics of mm-waves. The higher the frequency the shorter distance a wave propagates unless it gets a serious power boost. But the heterogeneous network by definition will be composed of densely packed cells in urban environments, meaning no mm-waves will have to travel far between hops.

The traditional uses of the mm-waves include radio navigation, space research, radio astronomy, earth exploration satellite, radar, military weapons and other applications. The backbone/backhaul networks (point to point network) for existing telecom network to connect base station to main switching centre (MSC), Local Multipoint Distribution System (LMDS), indoor WLAN, high capacity dense networks are also present in the mm-waves. The typical microwave backhaul bands are at 6.0 GHz, 11.0 GHz, 18.0 GHz, 23.0 GHz, and 38.0 GHz frequency bands.

The light use of mm-waves could be attributed to high attenuation and low penetration. At such high frequency, waves are more prone to rain and other atmospheric attenuation. The wavelength is in the order of millimeters, and rain drops are also of the same size. Rains absorb high frequency waves and make it difficult for propagation. However, the experimental results show that in heavy rain condition, attenuation is 1.4 dB and 1.6 dB for 200 meters distance at 28 GHz and 38 GHz, respectively [8]. The rain attenuations at 60 GHz for a rainfall rate of 50 mm/h, is approximately 18 dB/km [11]. A proper link design with slightly high transmit power may take care of rain attenuation.

Slight change in the position would affect the signal strength at the receiving end, due to which mm-waves are deeply affected by scattering, reflection and refraction. The root mean square (RMS) delay spread for mm-waves is of the order of few nano seconds, and it is high for are non-LOS (NLOS) links than (LOS) links [9]. Similarly, path loss exponent for NLOS links is higher than LOS links. Due to higher path loss and RMS delay spread, it is assumed that mm-waves are not suitable for (NLOS) links. However, these difficulties could be managed by using carrier aggregation, high order MIMO, steerable antenna, beam-forming techniques.

Recently, extensive measurements to understand the propagation characteristics for defining the radio channel have been carried out at 28 GHz in the dense urban areas of New York City and at 38 GHz cellular propagations measurements were conducted in Austin, Texas, at the University of Texas main campus. The measurements were conducted to know the details about angle of arrival (AoA), angle of departure (AoD), RMS delay spread, path loss, and building penetration and reflelction characteristics for the design of future mm-wave cellular systems. The propagation feasibility studies at 28 GHz and 38 GHz showed that propagation is feasible up to 200 meters of distance [6,10] in both the conditions, i.e., (LOS) and (NLOS) with transmit power of the order of 40–50 dBm in a difficult urban environment. This is size of micro cell in the urban areas.

The frequency bands around 60 GHz is best suited for pico and femto cell due to high data carrying capacity and small reuse distance due to strong oxygen absorption at the rate of 15 dB/Km. The usage in frequency bands around 60 GHz is highly sparse, which provides freedom to allocate a large bandwidth to every channel. Moreover, equipment can be made very compact due to the very small antenna size.

Much research work has been done for indoor channel characteriszation at 60 GHz band but a very few work has been done for outdoor characteriszation. In reference [12] measurements were carried out for narrowband CW for received power against separation distance in different environments mainly airport field, urban street and city tunnel. A channel sounder based on correlation has been used for the measurement for centre frequency of 59.0 GHz with a bandwidth of 200 MHz. A 90°-horn antenna was used at transmitting end and a biconical horn with an elevation beamwidth of 20°

was used at receiver in all the measurements. The measurement was carried out for path loss exponent and RMS delay spread. The result found that path loss exponent was between 2 and 2.5 for outdoor environment and RMS delay spread was lower than 20 ns. Result also included that multipath phenomenon was bad at parking garage due to large dimensions and smooth surface as compared to city streets and road tunnel, where multipath phenomenon was not much significant.

In [13] measurements were carried out at 55 GHz in city streets of London (UK) with moderate traffic density using a fixed transmitter and a mobile receiver, with link distances not greater than 400 m. The transmitter installed at 10 m above the ground level and receiver was mobile mounted over the roof of a car. The test signal was narrowband FM signal generated through Gunn oscillator and fed to a 25-dBi horn antenna. The result found that path loss exponent was 3.6 for a T-R separation of 400 m with LOS path and path loss exponent was 10.4 for same Tx-Rx separation in NLOS condition.

In order to understand the radio channel propagation characteristics, extensive propagation measurements in urban environment hasve been carried out long back at the campus of Delft University of Technology, Netherlands [14]. The measurements for frequency fading over 100 MHz bandwidth centered around 59.9 GHz were done almost exclusively in the time-domain by using network analyzsers and channel sounders. The block diagram of the measurement system used in reference [14] for the frequency-domain characteriszation of the radio-channel is shown in Figure 4.3.

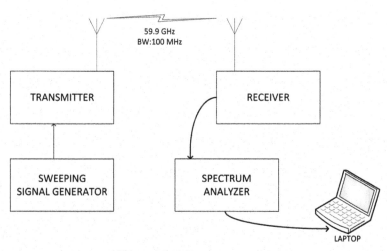

Figure 4.3 Measurement setup.

The two main components are signal generator towards transmitter side and spectrum analyser at receiver side. A flat omnidirectional antenna (2 dBi, 120°) was used at transmitter side and omnidirectional (120°) and patch directional antenna (pencil beam, 19.5 dBi, 15°) were used at receiver side. Measurements with both were done in order to see the difference in performance, because omnidirectional antenna allows for more reflected components to enter the receiver. The measurements were conducted for statistics of the 'k' factor of Rice distribution and the path loss coefficient for the pico cell of the order of 50 m radius at three different locations including outdoor and indoor. The measurements were done in possible locations for the mobile multimedia communication.

The measurements were taken in the corridor area (indoor) of the University for the Rice factor k and received power versus distance with T_X–R_X separation of 12–15 m are shown in Figures 4.4 and 4.5 below.

The measurements were taken in the parking area (outdoor) of the University for received power versus distance on logarithmic scale with T_X–R_X separation of 12–15 m is shown in Figure 4.6.

The measurement results show that propagation is feasible upto 10–15 m in the indoor and outdoor urban environment, which is normal size of pico cell.

The Radiocommunication Sector of International Telecom Union (ITU) is responsible for management of radio spectrum at international level. As per ITU-R frequency allocation plan [15], the frequency band 10–40 GHz has

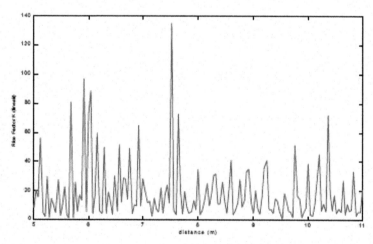

Figure 4.4 Rice factor k versus distance in the corridor. Directional receiver antenna used.

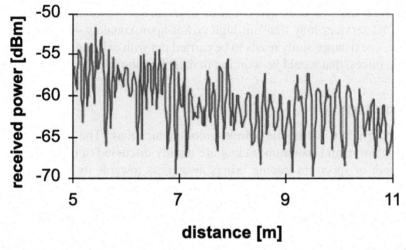

Figure 4.5 Broadband average received power in the corridor with omnidirectional receiver antenna used.

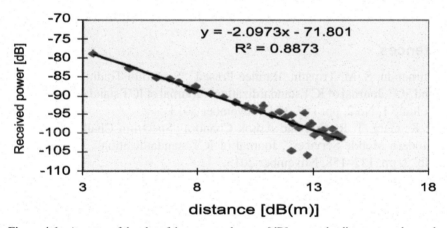

Figure 4.6 A scatter of the plot of the measured power [dB] versus the distance on a log scale for outdoor location (parking) with omnidirectional antenna.

been earmarked for satellite based services in all the three regions along with Fixed and mobile services. Local Multipoint Distribution System (LMDS), WLAN, Satellite services and High capacity dense network etc. are main services present in mm-waves. Several point to point fixed microwaves links are also working in this band. These links are basically for backbone/backhaul network for GSM and other services. A good amount of vacant spectrum is available at mm-waves which could be utiliszed for 5G communications services. 5G services may transmit high power approximately 40–50 dBW. Therefore, coexistence study needs to be carried out with existing LMDS and satellite services, that would be working in neighbouring spectrum bands.

4.6 Summary

Spectrum is a key to mobile wireless communications. The concept of spectrum sharing and spectrum trading are mainly discussed in this chapter. The concept of spectrum trading brings awareness towards its importance as it is not much successful in many countries. Spectrum management is an important challenge for the evolution of 5G communications. Extra high frequency bands have to be explored for the quick data transfers. Working on this, mm-waves have proved successful for the short-range communications. But still further research has to be carried out for the advancements in the existing technologies like cognitive networks, for the efficient use of spectrum.

References

[1] Purnendu. S. M. Tripathi, Ramjee Prasad, "Spectrum Trading in India and 5G", Journal of ICT standardization", Journal of ICT standardization, volume 1, No.2, pp. 159–174, November 2013.

[2] P. K. Garg, T. R. Dua and Ashok Chandra "Spectrum Challenges for Modern Mobile Services", Journal of ICT standardization", volume 1, No. 2, pp. 137–158, November 2013.

[3] Kwang-cheng and Ramjee Prasad, "Cognitive Radio Networks", John Wiley and Sons, Ltd, Publication, 2009.

[4] Cornelia-Ionela, Neeli Prasad, Victor Croitory, Ramjee Prasad, "5G based on Cognitive Radio", Wireless Personal Communications, volume 57, Issue 3, pp. 441–464, April 2011.

[5] Millimeter Waves. [online]. Available: http://www.ieeeghn.org/wiki/index.php/Millimeter_Waves

[6] R. Prasad and L. Vandendorpe, "An overview of millimeter wave indoor wireless communication systems," Universal Personal Communications, 1993. Personal Communications: Gateway to the 21st Century. Conference Record., 2nd International Conference on, vol. 2, no., pp. 885, 889 vol. 2, 12–15 Oct 1993 doi: 10.1109/ICUPC.1993.528506

[7] Rangan, S.; Rappaport, T. S.; Erkip, E., "Millimeter-Wave Cellular Wireless Networks: Potentials and Challenges," Proceedings of the IEEE, vol. 102, no. 3, pp. 366, 385, March 2014 doi: 10.1109/JPROC. 2014.2299397

[8] Ramjee Prasad, OFDM for Wireless Communications, Artech House, 2004.

[9] T. S. Rappaport et. al. Millimeter Wave Mobile Communications for 5G Cellular: It will work. Open Access, IEEE. (2013). DOI: 10.1109/ ACCESS.2013.2260813

[10] R. H. Ott and M. C. Thompson Jr., "Characteristics of a Radio Link in the 55 to 65 GHz Range", IEEE Transactions on Antennas and Propagation, pp. 873–877, 1976.

[11] S. Geng et al. "Millimeter-Wave Propagation Channel Characterisation for Short-Range Wireless Communication" IEEE Transaction on Vehicular Technology Vol. 58, No.1, January 2009.

[12] P. F. M. Smulders, L. M. Corria, "Characterisation of Propagation in 60 GHz Radio Channels," Electronics & Communication Engineering Journal, Vol. 9, No.2, pp. 73–80, Apr 1997.

[13] H. J. Thomas, R. S. Cole and G. L. Siqueira, "An Experimental Study of the Propagation of 55 GHz Millimeter Wave in an Urban Mobile Radio Environment", IEEE Transactions on Vehicular Technology, pp. 140–146, 1994.

[14] D. M Matic, H. Harada, R. Prasad, "Indoor and outdoor frequency measurements for mm-waves in the range of 60 GHz," Vehicular Technology Conference, 1998. VTC 98. 48th IEEE, vol.1, no., pp. 567, 571 vol.1, 18–21 May 1998. doi: 10.1109/VETEC.1998.686638.

[15] ITU-R Preparatory Studies for WRC-15 [online]. Available:www.itu.int/ en/ ITU-R

5

CYBER Security and Threats

The security challenges in communication networks are a major concern in today's world, with security threats and possible loopholes in communication systems appearing at much higher rate compared with technological advances in the communication systems in itself. Based on the operational structure of WISDOM 5G it is obvious to note the enormous security challenge that surrounds the overall operation of the network. Providing a unified access to the user and seamless migration between underlying access networks would necessitate enormous effort for securing the confidential data related to the user.

The security mechanisms to ensure reliable services utilized by the service providers rely on authentication, authorization, non-repudiation and confidentiality based mechanism among others. These mechanisms collectively ensure that service provider fulfil their liability to deliver reliable and trust worthy services. However, an underlying operational constraint of these mechanisms is the imposed latency. As WISDOM 5G requires almost zero latency for the user to access the services, it would be a major challenge to ensure reliable secure service and meet the objective of high data rate access to the customer in a ubiquitous manner.

Possible mechanisms that could be utilized for addressing the security challenges imposed on WISDOM 5G could be as follows:

Unique ID: Majority security mechanisms are initiated in networks to recognize a particular device in the overall network. If a given mobile device is recognized in a manner that differentiates it from other mobile devices then requirements for authentication and authorization could be avoided. Characteristics that could uniquely define a mobile device could be utilized for forming a unique ID.

Privacy by Design: Maintaining the data and user profile confidentiality is a major requirement imposed on the service providers through governing cyber laws. Security and privacy aspects are usually dealt with separate comparison, with the other network operations. Designing of network components and the overall operational backbone in WISDOM 5G with privacy and security aspects as prime objectives could ensure reliable privacy and security.

5.1 Major Challenges Surrounding Future Cyber Security

The plethora of challenges that surround cyber security can be broadly categorized into the following broad areas:

5.1.1 Network Borders

The emergence of 5G network as stated earlier that merges cells of different sizes and allows unified access technologies for the user device would lead to disappearance of network boundaries. Further the user device is expected to directly communicate with devices that comprise IoT for acquiring information about a certain physical parameter. This would be a different operational setup from today's communication network wherein government entities, private enterprises and private individual users could be the three broad network categories. The government entities currently impose strict operational rules regarding their communication network. Large private enterprises also lay down operational regulations regarding their communication network. These operational regulations would be hard to maintain if the networks are unified and user devices seamlessly switch between them [1].

5.1.2 Hindrance to E Commerce

The intention to provide ubiquitous computing and communication capability with data rate is to allow users to access rich and diverse information unhindered. That would be supported with unification of access technologies and underlying networks, along with support for high mobility of the user device. It can be anticipated in future that many companies would roll out novel information providing services that required unhindered data rates. This will, however, be hindered if the user is prohibited due to a certain networks' rule and regulation, or the data connection is interrupted due to requirements for authentication and authorization. Cloud and Internet based service provider are looking at opportunities of the form such as [2, 3]:

- Anything as a service (XaaS)
- Sensing as a service (SaaS)

Large enterprises that provide cloud based and Internet based services wish to utilize the aforesaid data service models for increasing the utility of Internet and unification of networks that would comprise machines (IoT, sensing as a service). However, the concept is imposing massive operational complexities even on the enterprises as they are unable to figure out the possible way to deliver such a plethora of utilities and create business opportunity.

Therefore, it can be safely presumed that if the data delivery model is tough to design and operate, security and privacy challenges that such a unified network would throw open would be much more complex to address.

The unified network operation promises to offer the end user the enormous utility of high data rate access with ubiquitous connectivity and extremely rich access to required information. The enterprises also seek to tap into the huge business potential that such a unified network could offer. However, a major privacy and security invasion issue for the common users could easily occur in such a densely connected network could lead to the governments imposing strong prohibitory orders that could ruin the economy and business significantly.

5.1.3 International Cyber Disputes

The unified network in WISDOM based 5G would depend significantly on an Internet based core network. Currently, there is a major debate prevalent on controlling the Internet. As it has enormous power in the form of millions of people social networking using it, or the possible spying and cyber crime committed using it. Committing cyber crime with an intention to ally to harm a nation or a major business entity directly, i.e., cyber war is discussed in the following sub-section. Many countries want to have a multilateral control on the Internet, i.e., countries would have direct control on the Internet activities concerning them, including data servers that are related with them but situated in another land; while some countries want a major international entity to take control of the Internet. In both the cases there are possible restrictions that would be forced on Internet operations in future. To protect sovereign interests an individual nation is likely to regulate the Internet autonomy, but this could have detrimental effect on the possible services foreseen to be fulfilled in future through 5G. The more connected the world through a unified network would be, the harder it would be to govern it and bring about an international consensus on Internet Governance.

5.1.4 Cyber War

In a unified mega network that allows users to access information directly from cyber physical structures (CPSs) that are in place for automation and control of a CPS, the CPS could be attacked/compromised to harm the possible operations derived from that physical structure. Such attacks on CPS have taken place in current communication network scenarios such as Heart bleed and Suxnet. Their likelihood would increase significantly if the networks are unified and users gain larger uninterrupted access. In the event of a large scale, cyber attack could lead to paralysis of critical infrastructure services such as transportation and banking and public utility such as electricity distribution. The 5G communication network would bring the physical infrastructure and the cyber infrastructure very close, thereby the threats to CPS would be enormous.

The governmental agencies that currently have a tough task in maintaining public utilities security and government infrastructure protection could encounter a situation of securing the infrastructure manifold complex than it is today. Cyber defence similar to other conventional defence mechanisms are cost intensive. Nations could be forced to increase their spending on cyber defence, in turn neglecting other critical priority issues that need to be addressed [1, 6].

5.1.5 Differentiation of Legitimate Versus Illegitimate Activities

The unification of various communication networks and devices, and the possibility of accessing information from devices diversely spread across the landscape would make it harder to determine as to what would comprise as legitimate and what would be illegitimate activities. Based on users' preferences and choices, the service provider wishes to mould that service and application to meet the users' requirement. This is a well-established practice today. In future, especially in the case of service delivery model such as sensing as a service and anything as a service, the service providers would try to data mine about the potential user/customer from their person specific data. The companies are highly interested in inferring the customer as closely as possible, since this influences their business prospects. As person specific data would grow based on the interaction a given person holds with the myriad of devices around, it would be very hard to draw a permissible line of what could be used for analytics and service providing and what comprises personal data that requires to be protected to ensure the privacy of the individual.

5.1.6 Legal System to Govern Machine-to-Machine Interactions

The legal systems around the world are designed keeping in mind the needs of the society, ensuring that by abiding by the law the whole society is secured. This is also applicable broadly for the cyber laws that have been enacted by the various governmental bodies. However, not in the 5G communication scenario that would comprise communication between machines (M2M) and IoT. The machines and devices would communicate with each other for certain self-automated tasks based on local decision making/intelligence, this could bypass the involvement of any human being governing the information exchange and automated decision making. In such a case a compromised machine may exploit a connected machine for gaining information or utilizing it to carry out an activity for illicit purpose. If the crime committed shifts from a human being to a machine, it would be tough to enforce any governing law existent today. The legal rules as applicable to common public are drafted such that they are comprehensible to a layman in the broad sense. For example a layman can understand reading Internet privacy and security rules as to what is the ethical way of using the Internet. However, drafting such a regulation base for machines/devices is infeasible considering the fact that interactions between machines would take place in a highly complex technical way, which could vary from machine-to-machine and underlying communication medium between the machines protecting network.

5.2 Users Awareness

For efficient user expertise of maintaining the expected cyber discipline in navigating across the plethora of data available it would be necessitated to educate the users of the usage etiquettes on the connected network. This is of tantamount importance as in a unified mega network in the form of WISDOM/GIMCV it would be necessary for the users to ensure that they access the data in an appropriate manner, following the basic guidelines. Some broad user guidelines and awareness are as described below:

The end users must be aware of these minimum safety measures [4]:

- To install the antivirus and anti-malware solutions in order to protect the devices from various virus or malware attacks. This idea has proved to be an effective solution against malicious attacks.
- The end users must be vigilant of the peculiar activities or behaviours in their own devices.

- They must install the applications from the original developer. The sources must be trustful having legitimate contact information and website.
- They should prefer to choose the applications on the Internet with the maximum number of downloads.
- They should not access their accounts or any other sensitive data when using those devices in public.
- They must ensure the frequent update of their operating systems and other software of their devices.
- They should also install a personal firewall to protect mobile device interfaces from direct attack and illegal access.
- The mobile network operators (MNO) should also be helpful to the end users by providing a secure environment. They can install antivirus and anti-malware software to scan outgoing and incoming SMS and MMS to the mobile network.

In addition to these, application developers should ensure that the sensitive or private data is not being sent to the unencrypted channel, which means data must be sent through HTTPS or TLS networks.

In brief we can note important players in securing the mobile wireless ecosystem. They include [3]:

- Mobile network operators (MNOs)
- Manufacturers of hardware, including mobile devices, chipsets and network equipments
- Application developers and market places
- Operating system vendors
- Network service providers
- Support software vendors
- Wi-Fi hotspot providers, over-the-top (OTP) providers and other platform providers

Most of the mobile industries are investing millions of dollars for providing the cyber security solutions for the strong security in the future. From the aspect of mobile operating systems security solutions include [4]:

- Mandatory encryption
- Application certificates
- Permission list for installations

The successful delivery of the mobile security or cyber security solutions require an active participation of the entire mobile landscape of mobile communication stakeholders, industry, Government, enterprises and finally the consumers.

5.3 Spectrum Related Security Issues in CRNs

As discussed earlier cognitive radio principles would be highly applicable to ensure reliable use of spectrum especially where the access would be relying on conventional cellular bands apart from the proposed use of mm bands. Security related issues that surround cognitive radio based networks are elaborated as below

The exceptional growth in the cognitive radio (CR) has attracted several researchers as this innovative concept has proved its potential worldwide. A CR network (CRN) should perform following functions [5]:

- To determine the portions of the available spectrum and detect the presence of licensed users when a user operates in a licensed band which is termed as *"Spectrum Sensing"*.
- To select the best available channel which is termed as *"Spectrum Management"*.
- To coordinate access to the channels with other users (secondary users) which is termed as *"Spectrum Sharing"*.
- To vacate the channel when the licensed user is detected which is known as *"Spectrum Mobility"*.

Some of the spectrum access related security issues concerned with CRNs are [7, 8]:

- *Masquerading of a cognitive radio node*: This threat identifies the masquerading of a CR node while collaborating with other CR nodes for CR functions: spectrum sensing, spectrum sharing, spectrum management and spectrum mobility. For example, a malicious device can send wrong spectrum sensing information to other CR nodes. The affected functionalities are spectrum sharing, spectrum sensing and spectrum mobility.
- *Selfish Misbehaviours*: During the channel negotiation process, a selfish cognitive node tries to gain an unfair advantage and try to improve its own performance. The channel negotiation process is done using the results from spectrum sensing and the fairness depends on the cooperation of the contending nodes. A selfish node may conceal the available data channels from others and reserve it for its own use. The affected functionalities in this case are spectrum sharing and spectrum mobility.

- *Hidden node problem:* This threat identifies the case when a CR node does not detect the user because of the obstacles. The consequence is that, it transmits the same frequency bands of the primary user causing harmful interference. The affected functionalities are spectrum sharing, spectrum mobility and spectrum sensing.
- *Jamming of the channel used to distribute cognitive messages*: This threat identifies the jamming of a cognitive control channel that is used to distribute cognitive messages in the CR network. This can be executed against an out-of-band or an in-band cognitive control channel if the frequency of the channel is known. The affected functionalities are spectrum sensing and spectrum sharing.
- *Unauthorized use of spectrum bands for Denial of Service to primary user:* This threat identifies the case where a malicious node or CRN emits power in unauthorized spectrum bands to cause Denial of Service (DoS) to primary users. The affected functionality in this case is spectrum sharing.
- *Malicious alteration of cognitive messages*: This threat identifies the alteration of cognitive messages that are exchanged in the CRN. The affected functionalities in this case are spectrum sharing and spectrum sensing.
- Eavesdropping: This is a common threat or problem in the wireless systems where the privacy of the data is communicated over the other systems. The eavesdropper may get the access to the exchanged content over wireless links like CRNs and then exploit the information against the network.

5.4 Summary

Addressing security and privacy challenges in respect to 5G would require excessive examination. In fact it can be concluded that reliable 5G operations would be unfeasible until suitable mechanism that would ensure dynamic operations with excessive high data rate and mobility could prevent cyber crime against user(s).

References

[1] ICSPA, Project 2020 Scenarios for the Future of Cybercrime–White Paper for Decision Makers [online], Available: http://2020.trendmicro .com/Project2020.pdf

[2] HP pursues 'sensing-as-service' [online], Available: http://www.eetimes.com/document.asp?doc_id=1257823

[3] Where the Cloud Meets Reality: Scaling to Succeed in New Business Models [online], Available: http://www.accenture.com/us-en/Pages/insight-cloud-meets-reality-scaling-succeed-new-business-models.aspx

[4] Jorja Wright, Maurice. E. Dawson Jr, Marwan Omar, "Cyber security and mobile threats: The need for antivirus applications for smart phones", Journal of Information systems technology and planning, volume 5, Issue 14, pp. 40–60, 2012.

[5] Abdullahi Arabo, Bernandi Pranggono, "Mobile Malware and Smart Device Security Trends, Challenges and Security", 19th International Conference on Control Systems and Computer Science, pp. 526–531, May 2013.

[6] "Today's mobile cyber security", Blueprint for the future by CTIA.

[7] Gianmarco Baldini, Taj Sturman, Abdur Rahim Biswas, Gyozo G´odor, "Security Aspects in Software Defined Radio and Cognitive Radio Networks: A Survey and A Way Ahead", IEEE Communicatons Surveys & Tutorials, volume 14, No. 2, 2012.

[8] Neeli Rashmi Prasad, "Secure Cognitive Networks", Proceedings of the 1st European Wireless Technology Conference, pp. 107–110, October 2008.

—

6

BEYOND 2020

5th Generation wireless systems is referred as beyond 2020 mobile communications which means the 5G standards can be introduced in early 2020s. Today's current advanced technology is LTE advanced, which provides a peak download speed of 1 Gbps and upload speed of 512 Mbps.

5G has been proposed in response to the needs of the connected society 2020 and beyond which means its standards are beyond the currently existing mobile broad band technologies like 4G/LTE and HSPA. It is also that 5G might solve the problem of frequency licensing and spectral management issues [1].

The current 4G LTE system uses advanced technologies such as OFDM, MIMO, Turbo Code, Hybrid ARQ, and sophisticated radio resource management algorithms. It is worth mentioning here that WISDOM together with 4G would lead to realization of 5G systems. The fundamental concepts of 5G are being evolved and developed from the existing technologies, and 5G systems by 2020 would fulfil the requirements of interconnected society by offering very high data rates, provision of transmitting large amounts of data, and security over the data. The 5G terminals have software designed radios; also it has different modulation schemes and error control schemes [1].

5G is a concept where personalization meets connectivity and networking technological innovations by integrating under one interoperable umbrella leading technologies, such as M2M communication technologies, cognitive radio and networking technologies, data mining, decision-making technologies, security and privacy protection technologies, cloud computing technologies, and advanced sensing and actuating technologies.

5G bundles multi-radio, multi-band air interfaces to support portability and nomadicity in ultra-high data rate communications using novel concepts, and cognitive technologies. The aim of 5G is to supersede the current propagation of core mobile networks with single worldwide core network. The objective is to offer seamless multimedia services to users accessing all IP-based infrastructure through heterogeneous access technologies [1].

6.1 Challenges for 2020 and Beyond

There are more than 5 billion wireless connected mobile devices in operation in today's world, most of which are handheld terminals and mobile broadband devices in portable laptops, tablets and computers. The result is the tremendous rise in the overall traffic on the wireless communication systems fuelled primarily by the uptake in the mobile-broadband. By 2020 and beyond the wireless devices would be approximately 7 trillion and therefore the traffic will rise manifold as compared with today. 5G system should be an intelligent technology capable of interconnecting the entire world without limits.

The cost of deploying, operation and managing 5G systems for many applications are also the major challenges as the services should be accessible to every common man living in the world [2].

Increasing bit rates led to increased energy consumption in BS. Main challenge for future mobile networks is to reduce power consumption. In cellular networks BS consumes more than 60% of the power so it is preferable to reduce power consumption in BS elements. Recent networks are designed with the consideration of high peak load, not with medium or low load consideration. In real scenario if the load on a network increased, then coverage will decrease and vice versa. To solve this power consumption issue, network topology will be designed in such a way that as load decrease BS starts to cover more regions, and some of the BSs can be shut down [3].

In the recent times, new types of devices, their respective applications and services are being developed and will appear to us in the future. We humans are connected to them through the mobile networks in our daily life. For example, medical devices, traffic lights, vehicles, etc. 5G should provide long term, efficient, high-performing solutions for all these types of services.

Development of 5G is not just replacing the current existing technologies, but it is the matter of evolving and complementing these technologies with new Radio Access technologies with respect to specific scenarios and user cases [2].

6.2 Future Mobile Technologies

Some of the future mobile technologies to be discussed are [3]:

- CR
- Beam Division Multiple Access

- Flat IP Support
- Support IPv6
- Pervasive Network
- Multi homing
- Group Cooperative Relay Technique
- Mobile Cloud Computing Support

6.2.1 Cognitive Radio

CR technology will be more efficient radio communications systems to be developed. This new radio technology share the same spectrum efficiently by finding unused spectrum and adapting the transmission scheme to the requirements of the technologies currently sharing the spectrum. CR will have knowledge of free channel and occupied channel, type of data to be transmitted, modulation scheme, position of receiving equipment and also aware of the environment. With the knowledge of above parameters radio should capture the best available spectrum to meet user requirements and quality of services [2].

When the level of occupancy increases then these systems have to move continuously from one channel to another which reduces the efficiency of the system. As use of CR increases a single frequency, signal will appear on a new frequency continuously so that effective algorithm must be developed and CR system will move only when it is necessary [3].

A. Continuous spectrum sensing
In this system spectrum occupancy will monitor continuously, and CR system will use the spectrum on a non-interference basis for the user.

B. Monitor for empty alternative spectrum
When primary user returns to the spectrum, then CR system must have an alternative spectrum available, so that it can switch to secondary user on it.

C. Monitor type of transmission
The CR must have knowledge of transmission used by users so that interference can be ignored.

6.2.2 Beam Division Multiple Access

FDMA, TDMA, CDMA and OFDM are the various multiple access techniques used in the wireless communications. In these frequency and time are divided among multiple users. But Korea has proposed a new technique called BDMA which is known as "Beam Division Multiple Access" as radio interference for 5G which does not depend on frequency/time resources.

In BDMA technique BS allocates separate beam to each mobile station and it divides the antenna beam according to the location of mobile stations. Based on the moving speed and position of mobile station, the calculations of direction and width of a downlink beam are done by the BS. When mobile stations are located at different angles with the BS then BS transmits different beams at the different angles to transmit data simultaneously. If mobile stations are at the same angle with the BS, they share same beam. This multiple access technique significantly increases the capacity of the system [4].

The BS can change the width of beams, number of beams and direction according to a communication environment. When the mobile station and BS know each other's position or when are in LOS , they will communicate with each other by a separate beam. Maximization of radiation efficiency of antenna can be done by matching the radiation pattern of mobile station and BS antennas [4].

6.2.3 Support IPv6

In the 5G system, each mobile phone will have permanent "Home" IP address and "care of address" which represents its actual location. If computer on the Internet wants to communicate with cell phone, it sends a packet to the home address and subsequently server on home address sends a packet to the actual location through the tunnel. Server also sends a packet to the computer to inform the correct address so that future packets will be sent on that address.

Because of this IPv6 has to be used for mobility. IPv6 addresses are 128 bit, which is four times more than 32 bit IPv4 address. This 128 bit address will be divided into four parts.

- The part represents the home address of a device.
- The second 32 bits may be used for care of address.
- The set of 32 bits used for tunnelling to establish a connection between wire-line and wireless network.
- The last 32 bits used for IPv6 address may be used for VPN sharing.

From this we can conclude that future 5G technology has a goal of establishing single worldwide network standard based on IPv6 for control, packet data, video and voice. The users can experience uniform voice, video and data services based on IPv6 [3].

6.2.4 Flat IP control

In 5G world, it is beneficial to transmit all voice, video and data using packet switching instead of circuit switching. It is an important feature to make 5G acceptable for all kinds of technologies. Each mobile device has to be allocated IP based on connected network and its location; devices are identified using the symbolic name instead of conventional IP format in Flat IP. This means the data is no longer routed by traversing a hierarchy from originating user through multiple layers of aggregation to a central core and then re-routed back out in a multilayer disaggregation hierarchy to the targeted user. The flat core routers of the originating user routes data directly to the local flat core router of the targeted user. In this technology, only one access specific node type is available.

This technology uses the reduced number of components lowering the operation cost and investment. This is the reason when there occur low system failure and latency. The only issue is the security as the Internet is open for hackers and criminals along with developers. Trojan horses and phishing are the two important security challenges [3].

6.2.5 Multi Homing

It is a technique used to increase the reliability of the Internet connection for an IP network. Future generation networks will support vertical handover, and user can simultaneously be connected to several wireless access technologies and move between them. The Internet through multiple network interface or IP addresses accessible to a single device is called multi homing. The configuration of this network assigns multiple IP addresses to different wireless technologies available on the same device. If one of the links fails, then its IP address will be unreachable but other IP address will still work so we can access the Internet.

Multi homing has become popular because of IPv6 address availability, that support more network protocol for multi homing than traditional IPv4 address. This IPv6 address has availability of provider independent address space. This technique works like IPv4, supports traffic balance across multiple providers and maintains existing TCP (Transmission Control Protocol) and UDP (User Datagram Protocol) sessions through cut-overs [3].

6.2.6 Pervasive Networks

The growth in the mobile broadband technology increases air interference technology and provides local area connectivity to the wide area. Future network will be "network of networks" which will provide uninterrupted service when roaming across many radio access schemes. The user can simultaneously be connected to several wireless access technologies and move between them which can be 2.5G, 3G, 4G or 5G mobile networks, Wi-Fi, WPAN, or any other future access technology. In 5G user can provide multiple concurrent data transfer and also user can move globally. Beyond 4G, network gives media independent handover, vertical handover and IEEE 802.21 support.

This IEEE standard 802.21 supports handover between same type of networks as well as distinct type of networks. Mobile IP provides vertical handover mechanisms for different types of networks, but can be slow in the process. To support this mechanism, mobile station must have dual mode cards so that it can work on WLAN and UMTS band and modulation scheme. To support vertical handover mobile station must have dual mode cards so that it can work on WLAN and UMTS band and modulation scheme [3].

6.2.7 Group Cooperative Relay Techniques

With the development of the MIMO systems, there is also higher throughput and reliability in a wireless network. This technology seems beneficial for BS side, but not on the user side due to size and power consumption. The alternate solution for this is group cooperative diversity techniques. In cooperative communication, source transmits data to the destination and at that time neighbour user (relay user) can also hear a transmission. The relay user also processes and forwards this message to the destination where received signals are combined. Both signals are transmitted with the different path as this gives diversity in the relaying the information. [3].

6.2.8 Mobile Cloud Computing Support

Cloud computing is a new and unique technique to access data like documents, application, video files, music file, etc., from any place without carrying any data-storage devices. A cloud user can access all data from anywhere in the world any time, and the best example of cloud computing is Gmail.

Mobile world is depended upon two factors:

- Network availability (3G, 4G, WIFI, etc.,)
- Handset availability (Feature phones, Smart phones).

The cloud computing is best option for low processing capability, low data storage. Some of the mobile applications that run on one mobile phone cannot operate on other phone. This problem can be solved by mobile computing. In this, the application runs on the specific device called cloud, and user can access the data and application. Requirement is user must have the Internet, without need for more computing capacity [3].

6.3 High Altitude Stratospheric Platform Station Systems

The transfer of the large data requires large band width. To use high-bandwidth solution is to use lower wavelength waves, which require LOS propagation, that are challenged to compare with lower frequency propagation. Wireless communication services are facilitated by terrestrial and satellite systems. The terrestrial systems are used to render services in complex propagation areas. Satellite links are used during the lack of availability of terrestrial links. Nowadays research is going on aerial platforms at high altitudes to provide LOS propagation.

A High Altitude Platform (HAP) is powered by battery, engine or solar cell. HAPs work similar to a BS and can be compared with a tall antenna that delivers a wireless communication. HAP is powered by battery, engine or solar cell. HAP can be easily deployed in hours which make it favourable in emergencies and disasters.

HAP does not require expensive launching like satellites which gives cost-effective way. HAPs will vary in position vertically and laterally depending upon the wind. This movement changes the look angle from the ground terminal. If this variation is greater than beam width of an antenna, then it requires the gain to operate the link.

HAPs provide a coverage radius of 30 Km, giving us the benefit of establishing single HAP instead of several terrestrial BSs in suburban and rural areas [4].

6.4 Human Bond Communications

WISDOM based 5G communication would usher the society into a digital age whereby humans can communicate in a real sense. The conventional communication network carries information between users in a 'digitized' form, which is significantly different from the real communication that is anticipated between two users if they are physically present next to each other. Human beings interact with the environment through five senses, i.e., skin, eyes, ears, nose and tongue. Overall understanding of a physical subject by human beings depends on the degree to which the observations are made by the five senses. However, the conventional information communication mechanisms have been centred on either visual (eyes) or audio (ears) based communication. Therefore, for a particular subject the overall judgement that can be drawn by human beings would be based on just these two senses, ignoring the possible utility of other three sensory systems completely. Accordingly, the judgement by the brain based on just the two sensory inputs is indeed only a partial representation of the actual physical subject. Human bond communication (HBC) stresses on utilizing all the five senses for modelling a physical subject with appropriate representation in digital form based on appropriate actuation and transmission across the communication network to support all sensory information [5]. WISDOM based 5G network that can support very high data rate would be capable of supporting the communication bandwidth requirements, if the overall sensor data is to be transmitted. Forming an almost true (complete) understanding of a physical subject through HBC supported on WISDOM 5G would create an information rich society that is unimaginable in today's communication world.

6.5 CONASENSE – Communication, Navigation, Sensing and Services

The WISDOM based 5G amply covers the communication aspect of the network, which as discussed earlier would involve one large core network. It has also been stated the significance of sensing the environment is through the capacity of individuals to gain rich information from machines (IoT). However, one of the objectives is to allow seamless connectivity to users even if they are mobile at high speeds. In the case of a user who is mobile at high speeds, it is just not the high speed connectivity that is necessitated, providing the user with effective navigational information is also an expected service requirement.

CONASENSE stresses on collective addressing of communication, sensing and navigation aspects dependent on the context, as the three aspects bidirectionally relate with each other to deliver services to the end user [6], [7]. The services that these three aspects provide have been referred as Quality of Life services. This justifies that QoL (Quality of Life) for common people would be ensured in future only if these aspects are collectively handled. This is validated considering the enormous utility of intelligent transportation systems (ITS) and V2V. Both the broad areas rely heavily on navigation information and both in turn influence the information and communication that is available to the end user.

HBC and CONASENSE are complementary aspects in terms of enormous utility of sensing obtained from the myriad types of physical sensors that could be deployed in the biosphere for garnering information. HBC stresses on specific sensing capabilities that human beings utilize to interact with their environment. Therefore HBC and CONASENSE would collectively ensure that information requirements of the future society in 2020 and beyond are appropriately met. The relation between WISDOM, HBC and CONASENSE, and their collective result as assured through the advent of 5G is shown in Figure 6.1.

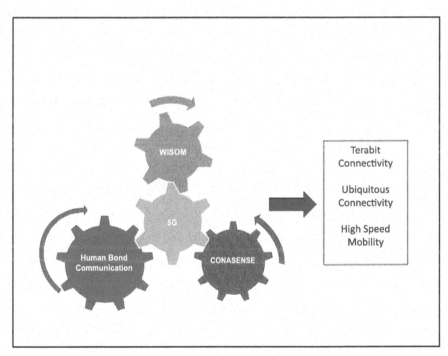

Figure 6.1   WISDOM, HBC and CONASENSE Collectively.

6.6 Summary

5G technologies will play a vital role in our lives by enabling unlimited access to information and data sharing to anyone, anywhere at any time. To fulfil the requirements of the users to experience Terabit communications, a 5G system must be a combination of different technologies like integrated Radio Access Technology (RAT), including evolved versions of LTE and HSPA and other special advanced technologies.

References

[1] Ramjee Prasad, "Global ICT Standardisation Forum for India (GISFI) and 5G Standardization", Journal of ICT Standardization, volume 1-No. 2, pp. 123–136, November 2013.

[2] Cornelia-Ionela, Neeli Prasad, Victor Croitory, Ramjee Prasad, "5G based on Cognitive Radio", Wireless Personal Communications, volume 57, Issue 3, pp. 441–464, April 2011.

[3] Ericsson White Paper, "5G Radio Access" http://www.ericsson.com/res/thecompany/docs/publications/ericsson_review/2014/er-5g-radio-access.pdf

[4] Saurabh Patel, Malhar Chauhan, Kinjal Kapadiya, "5G: Future Mobile Technology-Vision 2020", International Journal of Computer Applications (0975–8887) Volume 54– No.17, September 2012.

[5] Ramjee Prasad, Human – Bond Wireless Communications, Wireless World Research Forum, May 20, 2014, Marrakech, Morocco.

[6] Leo P. Ligthart and Ramjee Prasad, "CONASENSE – Communications, Navigation, Sensing and Services", River Publishers, 2013.

[7] Ernestina Cianca, Mauro De Sanctis, Albena Mihovska, Ramjee Prasad, "CONASENSE: Vision, Motivation and Scope", Journal of Communication, Navigation, Sensing and Services (CONASENSE), vol 1, issue 1, January 2014.

Index

About the Author

Ramjee Prasad is currently the Director of the Center for TeleInFrastruktur (CTIF) at Aalborg University, Denmark and Professor, Wireless Information Multimedia Communication Chair. Ramjee Prasad is the Founder Chairman of the Global ICT Standardisation Forum for India (GISFI: www.gisfi.org) established in 2009. GISFI has the purpose of increasing the collaboration between European, Indian, Japanese, North-American and other worldwide standardization activities in the area of Information and Communication Technology (ICT) and related application areas. He was the Founder Chairman of the HERMES Partnership – a network of leading independent European research centres established in 1997, of which he is now the Honorary Chair.

He is a Fellow of the Institute of Electrical and Electronic Engineers (IEEE), USA, the Institution of Electronics and Telecommunications Engineers (IETE), India, the Institution of Engineering and Technology (IET), UK, Wireless World Research Forum (WWRF) and a member of the Netherlands Electronics and Radio Society (NERG), and the Danish Engineering Society (IDA).

He is also a Knight ("Ridder") of the Order of Dannebrog (2010), a distinguished award by the Queen of Denmark.

He has received several international awards, the latest being 2014 IEEE AESS Outstanding Organizational Leadership Award for: *"Organizational Leadership in developing and globalizing the CTIF (Center for TeleInFrastruktur) Research Network"*.

He is the founding editor-in-chief of the Springer International Journal on Wireless Personal Communications. He is a member of the editorial board of other renowned international journals including those of River Publishers. Ramjee Prasad is a member of the Steering committees of many renowned annual international conferences, such as, Wireless Personal Multimedia Communications Symposium (WPMC); Wireless VITAE and Global Wireless Summit (GWS). He has published more than 30 books, 900 plus journals and conferences publications, more than 15 patents, a sizeable amount of graduated PhD students (over 90) and an even larger number of graduated MSc students (over 200). Several of his students are today telecommunication leaders worldwide.

Lightning Source UK Ltd.
Milton Keynes UK
UKOW06n1146270117
293027UK00002B/2/P